NGO FIELD WORKERS IN BANGLADESH

To my wonderful father and mother

NGO Field Workers in Bangladesh

MOKBUL MORSHED AHMAD
University of Dhaka, Bangladesh

LONDON AND NEW YORK

First published 2002 by Ashgate Publishing

Reissued 2018 by Routledge
2 Park Square, Milton Park, Abingdon, Oxon OX14 4RN
711 Third Avenue, New York, NY 10017, USA

Routledge is an imprint of the Taylor & Francis Group, an informa business

Publisher's Note
The publisher has gone to great lengths to ensure the quality of this reprint but points out that some imperfections in the original copies may be apparent.

Disclaimer
The publisher has made every effort to trace copyright holders and welcomes correspondence from those they have been unable to contact.

A Library of Congress record exists under LC control number: 2002102829

ISBN 13: 978-1-138-71986-6 (hbk)
ISBN 13: 978-1-315-19525-4 (ebk)

Contents

List of Tables

Acknowledgements

First and foremost, I want to thank the supervisors of my doctoral thesis Dr. Janet Townsend and Dr. Peter Atkins, both at the Department of Geography, University of Durham, Durham, UK, for their support, suggestions and encouragement. Professor David Hulme helped greatly with my Conclusions.

Special thanks to the Association of Commonwealth Universities (ACU) and the UK Commonwealth Scholarship and Fellowship Commission for awarding me the scholarship to pursue this research and its field work.

My thanks are extended to Mr. Shafiul Alam, Mr. Derek De Silva and all other Staff of MCC, Dr. Quazi Farooq Ahmed, Mr Shahabuddin, Mr. Asgar Ali Sabri and all staff of PROSHIKA, Dr. Rezaul Hoque, Mr. Pranab Kumar and all staff of RDRS Bangladesh, Mr. Simon Mollison, Mrs. Lulu and Mr. Saiful Islam of SCF (UK) and directors and staff of SCF (UK) 'partners' for their generous help and co-operation.

I also wish to thank Dr. Syed Hashemi who was working with the Grameen Bank, Professor M. Assaduzzaman, Dr. Noor-e-Alam Siddiqui of the Department of Public Administration, Professor A. Q. M. Mahboob, Department of Geography and Environment, Dhaka University, Mrs. Rina Sen Gupta of Red Barnet (Danish Save The Children). In the UK, Dr. Anne Marie Goetz of IDS, Dr. Michael Edwards (now Ford Foundation), Dr Uma Kothari at the University of Manchester and Dr David Lewis of the Centre for Voluntary Organisations at the London School of Economics freely gave invaluable advice and co-operation.

Sincere thanks to my students who helped me in conducting my questionnaire survey. I acknowledge the co-operation of my friends and staff at the Department of Geography, University of Durham, UK.

Finally, I owe much gratitude to my parents and sisters for their support and encouragement.

List of Acronyms and Abbreviations

AC   Area Coordinator (PROSHIKA)
ADAB   Association of Development Agencies in Bangladesh
ADC   Area Development Centre (PROSHIKA)
ASA   Assistance for Social Advancement (NGO-Bangladesh)
ASOD   Assistance for Social Organisation and Development
ASSP   Agricultural Support Services Programme
ATM   Assistant Thana Manager (RDRS Bangladesh)
BRAC   Bangladesh Rural Advancement Committee (NGO)
BRDB   Bangladesh Road Development Board
CCDB   Christian Centre for the Development of Bangladesh
CDEC   Comprehensive Development Education Centre of RDRS
CWDS   Char Atra Women Development Society, Shariatpur
DANIDA   Danish International Development Agency
DC   Deputy Commissioner (GO-Bangladesh)
DC   District Coordinator (RDRS Bangladesh)
DEW   Development Education Worker (PROSHIKA)
DFID   Department for International Development (formerly ODA-UK)
EDW   Economic Development Worker (PROSHIKA)
EU   European Union
FBCCI   Federation of Bangladesh Chambers of Commerce and Industries
FDF   Farm Development Facilitator (MCC Bangladesh)
FW   Field Worker (SCF UK) 'partners')
FY   Financial Year
GB   Grameen Bank (Bangladesh)
GK   Ganosasthya Kendrea
GO   Government Organisation
GOB   Government of Bangladesh
GRO   Grass Roots Organisation
GSS   Gano Shahayajyo Shangstha
HDF   Homestead Development Facilitator (MCC Bangladesh)
HKI   Helen Keller International
HRDP   Homestead Resource Development Programme (MCC Bangladesh)
HRM   Human Resource Management

IMF	International Monetary Fund
MCC	Mennonite Central Committee
NAB	NGO Affairs Bureau
NCB	Nationalised Commercial Bank
NGDO	Non-Governmental Development Organisation
NGO	Non-Governmental Organisation
NNGO	Northern Non-Governmental Organisation
NPO	Nonprofit Organisation
ORT	Oral Rehydration Therapy
PDP	People's Development Programme
RDP	Rural Development Programme (BRAC)
RDRS	Rangpur Dinajpur Rural Service
RLF	Revolving Loan Fund
SCF (UK)	Save the Children
SDS	Shariatpur Development Society, Shariatpur
SDS	Social Development Society, Madhupur
SNGO	Southern Non-Governmental Organisation
TC	Training Coordinator (PROSHIKA)
TM	Thana Manager (RDRS Bangladesh)
TNO	Thana Executive Officer (GO-Bangladesh)
UO	Union Organiser (RDRS Bangladesh)
WB	The World Bank
WDC	Women Development Coordinator (MCC Bangladesh)
ZC	Zonal Coordinator (PROSHIKA)

Glossary

Chars: shoals in rivers.

Civil society: civil society, or 'civic space', in this book occupies the middle ground between state and the private sector, and between state and family.

GROs (Grassroots organisations): GROs are membership organisations. The most important difference between GROs and NGOs lies in their accountability structures: GROs are formally accountable to their members, while NGOs are not. GROs are also sometimes called 'Community Based Organisations' (CBOs).

New social movements (NSMs): are the loosest of these groupings, and are often not formally organised at all; they may rather be people seeking to change their daily lives in roughly the same direction, perhaps linked by personal contact.

NGO (Non-governmental organisations): any group or institution that is independent from government, and that has humanitarian or cooperative, rather than commercial, objectives. Specifically, NGOs in this book work in the areas of development, relief or environmental protection, or represent poor or vulnerable people.

Nonprofit organisations (NPOs): the basic characteristics of NPOs are: *organised*, i.e. institutionalised to some extent; *private,* i.e. institutionally separate from government; *non-profit-distributing,* i.e. not returning any profits generated to their owners or directors; *self-government,* i.e. equipped to control their own activities, and *voluntary,* i.e. involving some meaningful voluntary participation.

Third sector: this can be designated variously as the voluntary sector, or the nonprofit sector. The *first sector* is the state and the *second sector* is the market. The *third sector* depends more on voluntaristic mechanism, involving processes of bargaining, discussion, accommodation and persuasion. These are obviously ideal types, and they are not mutually exclusive.

Note   Third sector and nonprofit organisations (NPOs) have been used interchangeably in this book. Both are common terms in the North. The exchange rate in December 2001 was $1 US = Tk 57

Chapter 1

Introduction

The purpose of this book is to explore Bangladeshi NGOs (Non-Governmental Organisations) 'from below'. I worked with NGO field workers in Bangladesh, exploring their careers, personal and professional lives, interaction with clients and superiors, and their opinions on the policies and activities of their NGOs. This book seeks to promote a better and fairer use by NGOs of the hidden resource which their field workers represent. I aim to show that the field workers of NGOs, the people working directly with the clients, are an undervalued and underused resource.

On a worldwide scale, NGOs are gaining increased attention and resources from policy makers, donors, academics and others. 'Donors' are the sources of 'overseas aid' for countries like Bangladesh. They include multilateral bodies such as the World Bank and the International Monetary Fund, northern governments and private foundations. The donors accept NGOs as agents for social welfare alongside the state, and as capable of fostering democracy in the South.[1] This is even more the case in Bangladesh. Since establishing independence in 1971, Bangladesh has been a donor-dependent country, both financially and functionally. During the 1980s, donors started to turn away from a state country which had been the recipient of almost all 'development' aid and advice.

I come from a Bangladeshi urban middle-class family characterised by absentee land ownership in our village and a house in the capital Dhaka. My father (Maniruddin Ahmed) was a civil servant who served in different positions in the national directorate of rural co-operatives for 32 years (throughout my father's career). I, as his son, have seen the decline of the co-operative movement, the rise and fall of the famous two-tier Comilla co-operatives and the 'NGO takeover' in Bangladesh.

---

[1]The South is used in this book to refer to the World Bank's Lower Income Economies, excluding China (where NGOs are very few and most conditions very different). The global North is constituted by the Higher Income Economies.

While undertaking my first MSc in Geography at Dhaka University, I studied the income-generation programmes of the two largest NGOs in Bangladesh - BRAC and PROSHIKA (Ahmad, 1991). That research focused on changes in the economic conditions of the clients of NGOs in rural Bangladesh. During my second MSc in Rural and Regional Development Planning at the Asian Institute of Technology in Bangkok, I wrote a thesis on the non-formal education programmes of two NGOs - ASOD (Assistance for Social Organisation and Development) and VERC (Village Education Resource Centre) (Ahmad, 1993), based on fieldwork at Sitakunda (South-eastern Bangladesh) and Rangpur (North-western Bangladesh). During this fieldwork, I discovered how neglected field workers are. I also found that field workers must simply carry out any directions given by their superiors. They were never consulted about why children do not go to school or why parents do not send their children to school or do ask them to stop going to school. Field workers are required to ensure high enrolment (particularly of girls) and low dropout rates, but they were never consulted on how to achieve this. Field workers were seen only as implementers, yet they were telling me that to achieve success in the non-formal education of children it would be necessary to target the parents and to raise both their social awareness and their economic status. They also told me that 'development' is something more than what their NGOs perceive. Talking to NGO managers and donors, I discovered how little they know about the realities in the field. I shall never forget these discussions with NGO field workers (which were beyond the scope of my AIT research). But I pledged those frustrated field workers 'If fortune favours, I shall do research with field workers and convey their message to NGO managers, donors, academics and policy makers'.

In 1995, I took a bold decision to leave the civil service in order to join the University of Dhaka as a lecturer (of which I sometimes repent). Once at the university, I started trying to pursue higher studies at a North American or West European university. Wherever I applied, my research proposal was the same, to study the field workers of NGOs. While writing this book, I still remember the faces of those field workers of VERC (Village Education Resource Centre) and ASOD (Assistance for Social Organisation and Development), who very plainly told me that I would forget them and their problems in the luxuries of Bangkok and Dhaka. My present book is a modest attempt to tell those field workers that I have not forgotten them, even though I have not been able to keep contact with them over the years.

When I came to Britain in 1996 to start my PhD research, I carried out a massive library and on-line search for research on the field workers of

NGOs. I found almost no research on field workers (other than volunteers) either in the North or the South. In Bangladesh, the only work on NGO field workers is that of Goetz and her colleagues (1997, 1996, 1995, 1994) with the women field workers of NGOs and Governmental Organisations in Bangladesh. So, NGO field workers are neglected not only by their managers and policy makers but by researchers too. Yet within the NGOs with which I have worked there is a rich texture of life, work and relationships that invites research.

My research focuses on a sample of four types of NGOs in Bangladesh - international, national, regional and (small) local. Although Bangladesh seems a monotonous flood-plain, the study areas represent four geographical regions of the country: a coastal area; one of the two Pleistocene Terraces; the poorer, least industrialised north-west; and one of the *char* (sandbank) areas of the riverine country.

One conflict which is apparent is between men and women field workers. When looking through the newspaper advertisements for field workers or managers in NGOs in Bangladesh, you will see, 'women are encouraged to apply'. To many men field workers and managers this seems to be due mainly to donor preference, which directs the policies and activities of NGOs in Bangladesh. I have seen great discontent among men field workers about the increasing importance the NGOs place on recruiting women. At the same time, women field workers complained to me that their work is undermined by their men colleagues who fail to understand the disadvantages for women in a traditional Muslim society. For example, women field workers cannot work after dark, for safety reasons. One NGO field worker told me that when he asked about the increasing number of women staff in his NGO, his superiors advised him not to be vocal about it because they (the men field workers and managers) had got jobs in the NGOs because of the women staff. His superiors told him that one condition for getting funds from donors is that they recruit women. It is not clear to me how much donors know about the negative impacts of their much loved gender policy.

Throughout my PhD research, my supervisor had serious problems with my English. I am grateful for her generosity in making corrections to my poor English. Apart from my own limitations, one reason for this could be the poor standard of English tuition in Bangladesh's education system. Unfortunately, my father could not afford to send me to an expensive English-medium school in Dhaka. Also, my supervisor did not find me critical enough, and she tried to make me so. I would link this with our colonial and post independence socio-political structure under which we all suffer from the desire to become *shahibs* (bosses) and do not like to be questioned. Interestingly, my research subjects (the field workers) complained to me about the hierarchy in their organisations where they are at the bottom and are the least valued.

I shall now explain how my book is arranged. In Chapter Two I deal with the NGO debate, and their role in 'development' in Bangladesh.

Chapter Three sets the scene for my specific field research. I describe my study areas, the working methods of my study NGOs, and changes in these NGOs. Finally, I make an effort to compare the benefits and facilities provided by the different NGOs to their field workers.

In order to understand more about the field workers, Chapter Four explores their socio-economic background from my interview research. This includes their family background, educational qualifications, career stage, etc. I also explore how and why they became field workers in NGOs, particularly their present one. The backgrounds of field workers and mid-level managers are then compared and the mid-level managers' choice of field workers is discussed.

In Chapter Five, I examine the personal lives of the field workers. Unfortunately this Chapter is a litany of their personal problems. The main problems are:- fear of job insecurity, financial difficulties, accommodation problems, family dislocation and risk. These problems vary by gender, marital status and the age of field workers. The expenditure patterns and recreational lives of the field workers are also of interest.

The professional lives of the field workers constitute Chapter Six. Again, this is a litany of professional problems, including transfer, promotion, group formation, training and work loads. Some field workers face difficulties in working with Muslim communities who suspect them of being agents of 'Christian' organisations engaged in evangelism. The field workers' assessment of their status in their NGOs, and the perceptions men field workers have of their women colleagues, and vice versa, are also important. The Chapter makes an effort to identify the weaknesses of the field workers as well as their strengths, which I think deserve due appreciation. Finally, the life of a single field worker highlights the corruption in NGOs in Bangladesh.

Chapter Seven examines the interactions between field workers and their clients and between field workers and their immediate superiors. It discusses the client/field worker interface, the motivations of the field workers, and the clients' perceptions of the services of their NGOs. Field workers' opinions of their clients, i.e. with what type of clients they prefer to work, are explored, as are the services clients demand from their field workers, but which they either do not get or field workers cannot give. There are many difficulties between field workers and their superiors. Field workers' assessments of their achievements, failures and future plans are included in this Chapter. I ask, how much are the problems and opinions of clients given due consideration in the planning of NGOs?

Chapter Eight details field workers' opinions of their NGOs' activities and policies, and of 'development'. I think field workers should be the movers and shakers of their NGOs so they have every right to give their opinion on the activities and policies of their NGOs, particularly in the case of microcredit which seems to be taken as a cure-all by most NGOs in Bangladesh for all the problems of poverty. Similarly, these 'development' workers can give valuable opinions on 'development'.

In the last Chapter, I conclude the findings of my research. From these, I also make an effort to suggest policy recommendations for NGOs towards their field workers and clients. I make recommendations for donors, the state and NGOs in order to make the activities of NGOs more effective and useful for their clients.

Chapter 2

Donors, the State, NGOs and their Clients in Bangladesh

Introduction

There are probably more and larger NGOs in Bangladesh than in any other country of the same size in the world. This Chapter first outlines the overall situation of NGOs in Bangladesh, then sets out in detail their relations with donors, state and clients and finally makes proposals for improvements. ADAB (The Association of Development Agencies in Bangladesh) had a total membership of 886 NGOs in December 1997, of which 231 were central and 655 chapter (local) members (ADAB, 1998). But the ADAB Directory lists 1007 NGOs, including 376 non-member NGOs. The NGO Affairs Bureau (NAB) of the Government of Bangladesh (GOB), which approves all foreign grants to NGOs working in Bangladesh, released grants worth about US$250 million in financial year 1996-97 to 1,132 NGOs, of which 997 were local and 135 foreign (NGO Affairs Bureau, 1998). NGOs have mainly functioned to service the needs of the landless, usually with foreign donor funding as a counter-point to the state's efforts (Lewis, 1993).

In this book NGOs refer to non-profit development organisations, or NGDOs. Salamon and Anheier (1997) define a non-profit organisation as:

a) *Organised*, i.e., institutionalised to some extent.
b) *Private*, i.e., institutionally separate from the state.
c) *Non-profit-distributing*, i.e., not returning any profits generated to their owners or directors.
d) *Self-governing*, i.e., equipped to control their own activities.
e) *Voluntary*, i.e., involving some meaningful degree of voluntary participation, either in the actual conduct of the agency's activities or in the management of its affairs (Salamon and Anheier, 1997).

The World Bank usually refers to non-governmental organisations as any group or institution that is independent of any government, and that has humanitarian or co-operative, rather than commercial, objectives. Specifically, the Bank focuses on NGOs that work in the areas of development, relief or environmental protection, or that represent poor or vulnerable people (The World Bank, 1996). This definition of an NGO is used in this book.

Changes in the Ways NGOs Work

Many of the national NGOs in Bangladesh began life in the 1970s after Independence, with radical agendas based on political reflection on the deep-seated, structural causes of poverty. Many leading figures had been previously linked to student left-wing groups (often with a Maoist ideology). Many had become disillusioned with the failure of radical political parties after the independence war (some also with Beijing's lack of support for an independent Bangladesh). During the 1970s they chose instead to channel their social activism into NGOs, initially into post-war relief efforts, and later into mobilisation of the poor in order to confront the structural causes of poverty (Davis and McGregor, 2000).

Internationally, NGOs developed the target group approach under which groups in particular need are identified and targeted. This has allowed NGOs in Bangladesh to work successfully with the rural poor and provide input into a constituency generally bypassed by the state. This approach emphasised the centrality of landlessness to a 'development' strategy, and placed the needs of landless women increasingly to the forefront of its programmes. A second innovation, by NGOs in Bangladesh, was the prioritisation of non-land-based sources of income-generation for landless women, who had been substantially neglected by the state. In particular, these income-generation activities are important for the survival strategies of poor women, i.e. to both man and woman-headed households. This innovation led to a concentration of efforts into small-scale, home-based income-generating activities such as cattle and poultry-rearing, food processing, social forestry, apiculture and rural handicrafts. These were combined with the provision of microcredit, to which the landless had previously been denied access, except from local moneylenders at high cost.

NGO initiatives in establishing income-generating activities proved to be an effective alternative to top-down, state programmes of rural works, but the extremely low rates of return on such activities have caused many to question their long-term sustainability (Aminuzzaman, 1998). In fact, some NGOs in Bangladesh still reject the idea of providing credit for income-generation activities in favour of organising the landless to strengthen control over assets such as land, forests and water-bodies and strengthen their claims on government services. Many of the larger national NGOs continue to combine both approaches, arguing that there are important social benefits to income-generating activities over and above their direct value, particularly in the case of rural women (Lewis, 1993). Some NGOs have had success in promoting human rights, particularly women's rights. This has been accompanied by backlash from the local elite, religious leaders and organisations (Rafi and Chowdhury, 2000; Shehabuddin, 1999). But NGO relations with their clientele appear to have become increasingly credit oriented, and there are now more restrictive rules, which mitigate against the feasibility of participatory procedures. The likelihood, therefore, of NGOs facilitating empowerment of poor people seems to have diminished during their expansion (Davis and McGregor, 2000; Ebdon, 1995; Montgomery *et al.*, 1996).

A Critical Evaluation

Even the largest NGOs in Bangladesh taken together cover only a fraction of the population: some have estimated that they reach only 10 to 20 per cent of landless households (Hashemi, 1995). Although NGOs claim that their social programmes are open to all, Rahman and Razzaque (2000) found that social programmes are closely linked with credit programmes, which creates a problem for participation in the social programmes by the extremely poor. The poorest 50 per cent are the target population but NGOs like BRAC prefer slightly better-off clients from among them, as the less poor are more likely to repay their loans on time than the less well-endowed or assetless. Nearly 10 per cent of BRAC members are still widowed or divorced women, a highly vulnerable social category (Montgomery *et al.*, 1996). NGO impact on poverty reduction has also been minimal. Montgomery *et al.* (1996) found little evidence that BRAC's clientele are altering their structural position within the rural economy. Only limited efforts have been made to make NGO operations truly participatory. Clients are seldom allowed to make decisions on programmes or budgets, or even to participate in monitoring and evaluation. Their participation is limited to relatively inconsequential areas of decision-making (Fisher, 1997). The solidarity and strength of groups of poor people is overshadowed by, and dependent on, the presence of the NGO. Thus, clients who want access to state relief or who demand higher wages may do so less because of their own strengths than because of the power of the

NGO, which in some areas is more influential than the village landlord, the local contractor or the state functionary (Hashemi, 1995).

Rather than promoting self-reliance, the NGO presence reinforces patron-client relationships; NGOs either replace old patrons or collude with them (Hoque and Siddiquee, 1998; Fisher, 1997; Ebdon, 1995). Too much dependence, especially by the majority of the rural poor, has led to a situation of 'NGO take-over' in rural areas. Indigenous social institutions have become weaker and are gradually disappearing. In the NGO infested areas, rural people, especially the poorer sections, look for the intervention of the NGOs in any socio-economic issues. The Union Parishads[1] as local participatory bodies thus become a merely symbolic institution with no physical programme for socio-economic uplift of the rural poor. NGOs, through their effective delivery system, management and responsive programme packages, have practically sidelined the Union Parishads and made the rural poor more dependent upon their intervention (Aminuzzaman, 1998).

If the general NGO priority is to provide resources and opportunities to those without them, one would assume that it would be considered a waste of time and money to begin work where these are already available. However, this has not been the case in Bangladesh. Ebdon (1995) found NGOs competing with each other in some villages for the same clients. Such competitive behaviour contradicts the NGO philosophy of co-operation and coordination with the common interests of empowering the poor. Instead, it creates and perpetuates factions and conflicts at all levels (Ebdon, 1995).

NGOs in Bangladesh have not originated from Grass Root Organisations (GROs) in civil society. Rather it is NGO workers who set up grassroots groups, which clients then join to get microcredit and other services. Most Bangladeshi NGOs are totally dependent on foreign funds. The volume of foreign funds to NGOs in Bangladesh has been increasing over the years, and currently stands at just below 18 per cent of all foreign 'aid' to the country in financial year 1995-96. Donors increased their funding from 464 NGO projects in 1990-91 to 746 in 1996-97, a 1.6 times increase in 6 years. The total amount disbursed by donors in Bangladesh showed a 143 per cent increase over the period (NGO Affairs Bureau, 1998). However, the distribution of funds to NGOs is highly skewed. The top 15 NGOs accounted for 84 per cent of all funding allocated to NGOs in

---

[1] Union Parishads, or Union Councils, are the lowest local level body in rural Bangladesh composed of elected members. In other words, the Union is the smallest administrative unit in Bangladesh. Bangladesh has 4,451 'unions', with an average population of 25,000 (BBS, 1998).

1991-92, and 70 per cent in 1992-93 (Hashemi, 1995). NGO dependence on donor grants have kept the whole operation highly subsidised. For example, annual operational costs of BRAC's branch-level units are still more than three times their locally generated income (Montgomery *et al.,* 1996).

High levels of donor funding have had two major consequences. *First,* NGOs have become donor-dependent, not merely in terms of the funding that is essential to their existence, but also in terms of seeking donor assistance to legitimise their activities. One large NGO had to stop its operations due to non-availability of funds (*The Daily Star,* 1999a). *Second,* upward accountability to donors has skewed NGO activities towards donor-driven agendas for development rather than to indigenous priorities. In keeping with the system of patron-client relationships, NGOs are considered as new patrons bringing with them access to external resources (White, 1999).

Most NGOs in Bangladesh maintain a high level of secrecy about their documents, staff salary and budgets. This makes the concepts of 'participatory' 'grassroot' 'development' advocated by the NGOs somewhat illusory. NGO staff are not allowed to form trade unions. Recently there have been allegations of misuse of funds, gender discrimination and nepotism against GSS, a large NGO. A state and donor investigation found that the rural-level women workers of GSS were compelled to go on maternity leave without pay, while GSS bought land worth millions of Taka to build its headquarters in Dhaka (Kabir, 1999). During the long process of NGO development in Bangladesh, many NGOs have certainly empowered themselves with buildings, while empowerment of the poor beyond better services has been rather limited. I find another interesting aspect of NGO activity in Bangladesh is that they never call for movements to root out corruption. How can social justice be achieved without reducing the present level of corruption in Bangladesh? It seems to me a hallmark of NGO dependency on donors that they are reluctant to promote social change.

Not only the NGOs, but also the economy of Bangladesh is highly dependent on foreign 'aid'. This dependency is not only economic, but, I think, something more. The World Bank thinks the quality of social services and of development efforts in rural areas could be enhanced through increased co-operation between government and NGOs. NGOs are better equipped than the government to deliver certain types of services - for example supporting micro-enterprise in rural areas - because their small size and flexible organisation give them an advantage. In other cases competition between public entities and NGOs may be conducive to greater efficiency. Sometimes co-operation between Governmental Organisations and NGOs leads to optimal results. This has been particularly true in the area of non-formal delivery of primary education and mobilising society for the increase of literacy (World Bank, 1996b). The influence of the World Bank also forced the government to cooperate in the 'Integrated Nutrition Project'. During the planning stages, BRAC and Helen Keller International (an NGO) worked very closely with local Governmental Organisations officials as full and equal partners. The most important aspect of NGO participation was to be in the implementation of the project (World Bank, 1996c). But Lewis (1997) points out that NGO-state partnerships are dependent in character on and driven primarily by, resource priorities.

The World Bank seems very satisfied with the performance of NGOs in providing microcredit in Bangladesh. It thinks of NGO-based institutions as effective and efficient delivery vehicles for (a) overcoming the failure of the formal financial sector to provide financial services to the poor; and (b) reducing poverty and correcting gender inequality. Therefore, according to the Bank, scaling up of NGO-based microcredit programmes is fully warranted within the context of a comprehensive poverty alleviation strategy comprising other complementary investments (World Bank, 1996d). However, the Bank points out that the development of NGO programmes should proceed in a manner and at a pace that existing institutional and human capital capacity can sustain. It recommends that undue pressure to speed up the pace of expansion of successful programmes in Bangladesh should be resisted. With these caveats, the Bank suggests important steps for integrating NGOs with commercial financial markets to develop an appropriate supervision and regulatory framework for the financial operations of NGO sector; encourage large NGOs to establish themselves as banks; encourage 'wholesaling' credit to established NGOs; and use smaller NGOs as brokers, utilise NGOs to mobilise self-help savings groups etc. (World Bank, 1996e). The NGOs' problem in providing microcredit has already been mentioned. The World Bank should perhaps take note of the poor accessibility of microcredit to the vulnerable and realise that giving priority to credit makes other aspects of 'development' almost ineffectual.

Government-NGO relations in Bangladesh have moved through stages of indifference and ambivalence (Kalimullah, 2000; White, 1999). In order to remain neutral in the eyes of the powerful, large, influential NGOs like BRAC have shown a tendency to retain good relations with local notables and the state (Montgomery *et al.*, 1996; White, 1999). But Aminuzzaman (1998) found that there was practically no relationship between the Union Parishads (local councils, above) and the NGOs. NGOs tend to 'mistrust' the Union Parishads and maintain a distance from them. Union Parishads, on the other hand, have a kind of 'suspicion' about the role and motive of NGOs (Aminuzzaman, 1998).

NGOs in Bangladesh have increasingly become subject to question and criticism by the state, political parties, intellectuals and the public in general. Recently, NGO activities and expenditures came under fire in the National Parliament and other fora (*The Daily Star*, 1999b; NFB, 1999d). One member of Parliament (MP) alleged that some NGOs raised money on false promises of jobs and credit, but misappropriated it. Another MP claimed that some NGOs make loans at the high rate of 14 per cent and resort to 'inhuman torture' on debtors who fail to repay on time (News from Bangladesh, 1999a). The relevant minister gave a face-saving answer to all these allegations, but in reality there is poor control by the state on NGOs in Bangladesh. This was reiterated by the head of NAB (NGO Affairs Bureau) on another occasion (*News from Bangladesh*, 1999b).

Some NGOs are flourishing simultaneously as service-oriented organisations and as profit-oriented business organizations (Davis and McGregor, 2000). The state is also being deprived of tax by NGOs taking advantage of loopholes in the regulations. Some senior officials of certain NGOs have used loopholes to become affluent (the Government's Audit Report, 1992).

BRAC is currently alleged to be running successful businesses like a commercial organisation, contrary to its charitable, tax free status. BRAC's cold storage company, a marketing organisation (named 'Aarong'), real estate company, restaurant etc. are highly profitable. Recently BRAC has received state approval to open a commercial bank for microlending (*Financial Times*, 1998). BRAC is also said to be planning to open a private university. BRAC provides no accounts of its commercial organisations' incomes or expenditures to any state department. White (1999) points out that BRAC generates 31 per cent of its income from its businesses. Given their tax free status, the private sector sees BRAC businesses as unfair competition.

PROSHIKA (another large NGO), has developed a transport company of 28 buses at a cost of 30 million Taka. At a cost of 15 million Taka, it has also established a printing press and a garment industry and is investing five million Taka in a video library. Recently it has started an internet and software business. In some cases, service-oriented NGO projects are basically market-oriented, with the objective of earning profits through long term capital investment. BRAC's cold storage project (costing 70 million Taka) and Ganosastha Kendro's highly profitable clinic, university and medicine businesses are striking examples. Allegations have also been made against GSS, which used donors' funds to open a printing press and media business but is not audited or taxed as per rules of the state (Kabir, 1999). The Finance Minister stated in a seminar that most NGOs are engaged in banking thus violating the law (*The Independent*, 1999a).

Due to strong donor support for NGOs, the state has in the recent past had to scrap its own desire to withdraw the registration of a number of NGOs, and even to change the head of NAB (NGO Affairs Bureau), who had appeared tough with NGOs which indulged in irregularities (Hashemi, 1995). When NAB cancelled the registration of three NGOs for financial irregularities, the head of a diplomatic mission in Dhaka personally intervened, brought the issue to the attention of the Prime Minister's office and got the cancellation order withdrawn. This action created great dissatisfaction among officials in the Bureau (Hashemi, 1995).

NGOs are non-democratic institutions, often dominated or dictated to by one individual, and many have a serious ownership problem. As NGOs are heavily dependent on foreign resources, in the absence of accountability, the flow of money from the outside can make the NGOs corrupt, controversial and autocratic (Zarren, 1996). Despite the negative effects, NGOs are accountable to the donor countries rather than the state of Bangladesh (Islam, 1995) let alone their clients.

In reality, the state is unable to control the NGOs. The NGOs often work against the directions and decisions of the state. Weak administration on the one hand and strong national and international backing on the other encourage some NGOs to defy the state and to work according to their own whims. In the recent past, the registration of ADAB (Association of Development Agencies in Bangladesh) was cancelled by NAB (NGO Affairs Bureau), but reinstated within a few hours. Naturally, this was achieved by a powerful international lobby.

Conclusion

Many NGOs in Bangladesh took part in or supported some of the popular movements against autocratic rulers (1990) and demands for free and fair elections (1996). All these movements were fostered by major political parties, which obviously antagonised the government in power (White, 1999). This has resulted in the politicisation of NGOs (Davis and McGregor, 2000; Hoque and Siddiquee, 1998; Hashemi and Hassan, 1999; *Independent*, 1999b). NGOs in Bangladesh cannot remain fully accountable to the government while simultaneously launching a challenge to government power. In reality, the only way to counter the influence of government and donors is through increased reliance on the clients - the rural poor. Only through the development of a system of accountability to the poor could NGOs truly transform themselves into organisations of the poor. Only by becoming organisations of the poor could NGOs truly prepare for a sustained struggle for empowerment.

Chapter 3

The Study NGOs

This Chapter outlines the strategy and methodology that were used to address the research questions and to realise the research aims and objectives of this study, setting these in the context of the NGOs and the places in which the research was conducted. Since my research is with field workers, I shall try to compare the benefits they enjoyed, because it is important to know about the NGOs and the environments in which my research subjects operate. Both the policies of the NGOs and the physical conditions and the infrastructural facilities greatly influence the lives and work of NGO field workers. Unfortunately, the field workers have little or no power to influence them. I shall also try to sketch the changes that my study NGOs have gone through (mainly through the voices of the field workers). Both the positive and negative aspects of these changes will be elaborated in later Chapters.

Choice of Organisations

For the category of 'international NGO', I worked with MCC Bangladesh (Mennonite Central Committee). I originally approached Action Aid Bangladesh and MCC. Action Aid was interested in allowing me to do the research, but their field workers were agitating in Bhola, where Action Aid decided to stop its direct work and hand over its activities to a local NGO. The field workers were agitating against job losses and the policy changes of Action Aid. Bhola was the oldest working area of Action Aid in Bangladesh and Action Aid had wanted to send me there. (Later, when I was about to finish my field work, the field workers formed a new NGO in Bhola called COAST). This illustrated for me some NGO conflicts.

One 'large national NGO' of Bangladesh was selected for the study, PROSHIKA (PROSHIKA, A Centre for Human Development). Here 'large NGO' means those NGOs which employ hundreds of field workers (for example BRAC, PROSHIKA, ASA) and work with thousands of clients. For example BRAC has 19,000 full-time employees, 34,000 part-time teachers and 2.3 million clients (96 per cent women) (*Economist*, 1998)

For this category of NGO I approached BRAC, PROSHIKA and ASA. PROSHIKA was the most enthusiastic and co-operative from the very beginning while the others were less positive. So, I decided to work with PROSHIKA. For the category of 'regional NGO', RDRS (Rangpur Dinajpur Rural Service) was my first choice because it is one of the oldest regional NGOs in Bangladesh. I got permission to work with them easily. RDRS has been working in the north-west of Bangladesh since Independence in 1971.

In the local NGO category I chose the local 'partner' NGOs of SCF (UK) (Save the Children Fund UK). These were formed by the former SCF (UK) staff when SCF (UK) ceased direct work. For reader's convenience I shall henceforth refer to these as the 'SCF partners', omitting 'UK'.

My choice of NGOs was based not only the size or reputation of the NGOs, but also on accessibility for the research. The openness of the case study NGOs to investigation was of the first importance.

The Foundations of the Study

My methods of inquiry for the study were predominantly ethnographic and/or qualitative, with limited quantitative work. They included in-depth case studies, structured/semi-structured interviews, participant observation and informal interviews/discussions with selected field workers, NGO managers and clients, as well as documentary searches.

Semi-structured/informal interviews, life histories and the questionnaire survey were all conducted in Bangla, save for meetings with key British academics in the UK and with the Bangladesh Director of SCF (UK).

Interviews

Using checklists of points in informal conversation, I interviewed a total of 97 field workers of four NGOs, and key informants such as their clients, their supervisors and senior managers of these NGOs. For interviews with women field workers, another woman was always present. In Bangladesh, it proved impractical to tape-record interviews (save with the Director of SCF (UK)) as this produced discomfort in the subjects.

A small number of life histories were obtained from field workers and from long-term NGO clients. Individual testimonies gave me access to the views and experiences of more marginalised field workers and clients. The great strength of oral testimony is its ability to capture personal experience and individual perception, which proved complementary to other interviews.

Questionnaire

The survey, of 109 questionnaires, was conducted with a random sample of men and women field workers in each study area by field assistants from the Department of Geography and Environment, University of Dhaka, under my supervision. It was not possible to find women field assistants.

Other methods

Focus groups were attempted, but proved impractical with both field workers and clients in view of their extremely busy schedules and the high levels of latent rivalry and conflict. NGO statements and documents proved of great interest on policy, and field workers' reports demonstrated the weight of paperwork which they must produce.

MCC Bangladesh

The Mennonite Central Committee (MCC) is an international NGO. Begun in 1920, MCC is the relief and 'development' arm of the Mennonite and Brethren in Christ Churches of Canada and the United States, with over 1,000 volunteers. North Americans, Asians and Europeans are involved in programmes of Agriculture and Community Development, Job Creation, Emergency Relief, Education, Health and Peace Concerns in over 50 countries around the world. All volunteers in MCC Bangladesh are from the North, and do not work as field workers.

MCC first came to Bangladesh to assist survivors of the great tidal bore disaster of 1970, centred at Noakhali. MCC believes that the spirit of the people of Noakhali and their eagerness to participate with MCC in 'development' became the basis for continued work there. There are now three main foci for the MCC programmes in Bangladesh: Agricultural and Family Development, Employment Creation and Emergency Asistance (MCC, 1998). I worked at Noakhali-Lakshmipur on the MCC Agricultural Programme, the goal of which is to improve the quality of life of rural poor people. Here 'poor' people are 'those farm or farm labour families who cannot produce their year's need of food for the family from their own resources' (MCC, 1998). The major objectives of the MCC Agriculture Programme are to increase: the ability of the farmers to utilise their resources more productively, agricultural production, nutritional and health status, income and the involvement of women in 'development' and to enhance education in their families (MCC, 1998).

To achieve the above, the Agricultural Programme has the following sub-programmes and activities:

1. *Farm Family Development Programme*: This programme works with the extension of technologies to farmers through the one-to-one (individual) contact method.

2. *Homestead Resource Development Programme*: This programme works with similar target sectors but organises the farmers into groups of between 20-25 members. The groups are encouraged to start a group fund by making weekly saving deposits which is then invested in group projects. This programme works with about 1700 farm families. 70 per cent of the groups are of women.

3. *Partnership in Agricultural Research and Extension*: This programme works with other NGOs around the country assisting them to develop an agricultural component in their organisations so that their farmers can also benefit from MCC's experience.

4. *Soybean Programme*: This programme does not work with any particular socio-economic group. It is involved in extending the soybean in Bangladesh with the intention of establishing it as a viable crop, and promotes soybean utilisation activities in selected working areas. In co-operation with the PARE programme, it also assists other NGOs in extending soybeans to those NGOs' contact farmers.

5. *Inoculant Bio-fertiliser Project*: This project was started with the intention of producing soybean inoculant for the soybean extension programme. MCC (Mennonite Central Committee) thinks that the inoculant project will in time to come produce other pulse inoculants to help farmers in Bangladesh.

6. *New Life Seeds*: This was a business set up within the framework of the Agriculture Programme, but as a separate entity to cater to the seed needs of the soybean programme. NLS procures soybean seeds for sale to farmers during the two soybean growing seasons in Bangladesh. Although set up as a business, so that it can operate on the profit it generates, to date it has not been able to earn enough profit to run without financial assistance from outside.

7. *Agriculture Training Programme*: This is a support programme which looks after the training needs of the Agriculture Programme. It assists the programmes in arranging training for staff within and outside the country. The ATP also helps the programme specialists develop

training manuals and training techniques. It assists with programme meetings, reviews and publications.

MCC (Mennonite Central Committee) has many expatriate volunteers working for its Agriculture Programme. They usually come for around three years and their food, lodging and transport costs are borne by MCC. A major component of the FFDP is research, and senior staff in this programme are technical. The expatriate staff or volunteers are mainly technical (for example the Technical Officer of the Soybean Programme). All Programme Leaders are Bangladeshi nationals. MCC policy-makers think that the advantage of putting expatriates in research and technical support is that they are well connected with external libraries and have more interest in doing research. The Farm Development Facilitators, Women Development Co-ordinators, Homestead Development Facilitators and extensionists are the field workers of MCC with whom I worked. A major feature of the Homestead Development Facilitators is that they work with groups (usually 20-25 members). All other field workers work individually but they organise some meetings with members.

All MCC field staff are recruited through open advertisements and have free weekends. Only in MCC and SCF 'partner' NGOs did I find the field workers able to enjoy weekends. Usually MCC field workers leave their working area (where they have to live) on Thursday evening to meet their families and come back to their working area on Sunday morning. The management structure seemed bureaucratic, which was confirmed by many field workers.

Noakahali-Lakshmipur

I mainly worked with the field workers of MCC in the *char* areas of the Noakhali and Lakshmipur districts, south-east of Dhaka (these are unstable land of shoals and sandbanks). Noakhali and Lakshmipur are well connected with the two largest cities of Bangladesh - Dhaka and Chittagong. It took me five hours to get to Noakhali from Dhaka by bus. Noakhali-Lakshmipur are closer to Chittagong which takes around three hours to reach. Bangladesh is one of the most cyclone-prone regions of the world. The vulnerability of this region to cyclones has brought many donors here. Thanks to the donors, cyclone shelters have been built. Except at cyclone times, these shelters are used as primary schools and the ground floor is a wonderful place for storing newly harvested crops. Additionally, there is a commendable coastal green-belt project which will not only help in the forestation process in the region, but also in creating ecological balance. The *chars* of Noakhali and Lakshmipur are similar, with mainly unsurfaced roads and no electricity in most places. There are many villages whose men are working abroad, mainly in the USA.

Noakhali-Lakshmipur is probably the most traditional region in the country. Women usually come out in a 'burkha' (veil) and a newcomer to any village should contact the senior men in the village. A new woman (say an NGO field worker) is expected to come in a veil.

An important aspect of MCC (Mennonite Central Committee) activity in Noakhali-Lakshmipur is that they are trying to popularise soybean among the farmers. Another important aspect of MCC activity is the research on agriculture. This is really interesting, as most NGOs in Bangladesh are involved in direct 'development' work. Another interesting thing about its activities is that the MCC has not joined the latest fashion for 'development' (microcredit), which it is really important to investigate (MCC members can borrow money from their groups' savings). A senior manager in MCC told me that he thinks microcredit is exploitative. All this is mainly due to the missionary nature of this NGO. Missionary NGOs are a) better funded b) more aware of religious sensitivities and c) more caring towards their workers. That is, the churches and private individuals who support MCC do so for its mission, while the donors of other NGOs want more specific results.

PROSHIKA A Centre for Human Development[1]

PROSHIKA is one of the largest national NGOs. I worked with the field workers of PROSHIKA at its Sakhipur Area Development Centre office. Sakhipur is a thana (subdistrict) in the Tangail District (north-west of Dhaka). PROSHIKA has 115 such centres comprising 156 administrative thanas of Bangladesh (PROSHIKA, 1997).

Since its inception in 1976, PROSHIKA's effort has been to engender a participatory process of 'development', and it claims to have succeeded in pioneering an approach that puts human development at the centre. This process is founded upon the understanding that poverty reduction and promotion of sustainable 'development' are dependent upon the human and material capacity-building of the poor, and their socio-economic and cultural empowerment through a process of generating human, social, economic and cultural capital. The very word PROSHIKA is an acronym of

---

[1] Please note that this is the correct name of the NGO, although I use the shortened form 'PROSHIKA'. Many PROSHIKA staff feel that outsiders diminish the NGO by using the shortened form, but the correct name is very long.

three words (the Bangla for training, education and action) which encapsulate the organisation's 'development' ethos.

Concurrent with group organisation and training, the poor are also encouraged to pool their resources to pursue employment and income generating projects. PROSHIKA has instituted a system of 'Revolving Loan Funds' to provide needed initial investment to those groups deemed ready to undertake these projects. In addition to initial investment capital, PROSHIKA also provides the necessary technical support to help make these projects succeed (Wong *et al*, 1998).

PROSHIKA works in 10,166 villages and in 654 urban slums in 45 districts. It works with nearly 1.3 million men and women clients drawn from poor rural and urban households and organised into 68,897 groups. This translates into a total programme reach of over 7.1 million individuals (PROSHIKA, 1997). The main areas of PROSHIKA activities are:

- Organisation building among the poor.
- Development education, including human development training, practical skills development training and people's theatre.
- Employment and income-generating activities.
- Environmental protection and regeneration.
- Universal education programmes.
- Health education and infrastructure programmes.
- Integrated multisectoral women's development programmes.
- Urban poor development programmes.
- Housing programmes.
- Disaster management programmes.
- Development policy analysis and advocacy.

The Management Structure

Although PROSHIKA says that it has a democratic management system and has no system of bosses, it was found in this study that there is a hierarchy in the management of PROSHIKA. Although all field workers call each other and their superior 'bhai' (brother) (compare White, 1999), the social structure in Bangladesh is hierarchical and so, I observed, is PROSHIKA.

Facilities and Benefits Enjoyed by Field Level Staff

All field workers get logistic equipment like motorcycles, and PROSHIKA pays for the fuel, with a ceiling each month of Tk 800. Most field workers I interviewed complained that this is not enough. PROSHIKA also pays for the maintenance of the motorcycles. There is no hire-purchase system in PROSHIKA like that in RDRS (see below). Many field workers reported that PROSHIKA's recent purchase of Indian-made motorcycles was a mistake since they have high maintenance costs, even though they are cheaper than the Japanese-made motorcycles. Women field workers are usually given 50cc. motorcycles but many try to avoid riding them due to shyness. Some even produce a medical certificate to prove that they are physically unable to ride a motorcycle, a familiar response among traditional Muslims in South Asia.

During my survey, most field workers complained that there is very limited scope for promotion in PROSHIKA, particularly for field workers. PROSHIKA has a different system of promotion for its field workers which I have not seen or heard of in any other NGOs in Bangladesh. A change of pay scale or elevation from one post to another one does not require any examinations, as promotions are on the basis of evaluation of the field workers by the Area Co-ordinator. But for a change of duty (for example for promotion from Economic Development Worker to Training Coordinator) all field workers have to sit a two-hour written test and face an interview. Some field workers complained of irregularities in the promotion and posting system in PROSHIKA. Earlier, there was no formal evaluation system in PROSHIKA, and the current system of evaluation was only introduced in 1996 after many years of demands. In the case of Economic Workers, the most important criterion for evaluation is the repayment rate of their credit. But still many field workers complained to me about irregularities by their immediate superiors in their evaluation.

All field workers get a house-rent allowance of 65 per cent of their basic salary per month at the thana level. They also get a medical allowance - special (for severe ailments) and general (two per cent of the basic salary). To get the 'general medical allowance' they have to submit their prescriptions as evidence. Most field workers thought that their general medical allowance should be given to them with their salary, as for government employees in Bangladesh. All PROSHIKA staff have their own provident fund. Like all staff, field workers can take loans against the amount deposited in their provident fund. Again, many field workers complained to me that there are irregularities in sanctioning loans. One field worker said that he applied for three consecutive years for a house-building loan. In the first year his application was rejected on the grounds that his evaluation was not good. In the second year his application was rejected because he hadn't enough savings in his provident fund. In the third year his application was rejected on the pretext that the money for that year had been exhausted and he had applied late. Most field workers complained that the staff at head office get loans more easily due to being on the spot.

All PROSHIKA field staff have to undertake certain residential training at Manikganj, one hour by bus from Dhaka. Most field workers like the training programme but try to avoid training because it hampers their field work. Also PROSHIKA charges its staff Tk 300 per course. When the field workers are on courses, their collection and distribution of money is hampered and they fear that this may affect their evaluation. Most training programmes last two to three weeks. The recruitment of women field workers has been a recent phenomenon in PROSHIKA.

Sakhipur Thana

Sakhipur is situated in the Madhupur tract, which is one of the two Pleistocene Terraces of Bangladesh. A major physiographic feature of the area is that it is on average 20 to 40 feet above sea-level, so it is flood-free.

Sakhipur had a major part of one of the largest forests of Bangladesh, but increased pressure on land for housing and agriculture, accompanied by the corruption of local administration and forest officials, has greatly reduced the forested area. Due to its poor agricultural potential, this thana still has a low population density, but recent improvements in irrigation technology have greatly helped farmers to grow IRRI (high-yielding paddy). At the same time, Sakhipur is famous for huge amounts of high quality bananas, guava, jackfruit and other seasonal fruits. Now farmers are showing an interest in growing vegetables. A major advantage for Sakhipur farmers is that it is well connected by roads with Dhaka, which takes a maximum of three hours to reach. Another major feature of Sakhipur is that many men from these villages work in South-East Asia (mainly in Singapore), so that their remittances they sent have a significant impact on the lives of people in Sakhipur. As one PROSHIKA field worker who has been working in Sakhipur for eleven years told me:

> When I came to Sakhipur eleven years ago, there were only two to three noticeable shops in the Sakhipur bazaar (market). There was no pucca (brick) house in Sakhipur, most of them were mud-walled. Now you see how big the Sakhipur bazaar is. Now you can see many brick made houses in remote villages.

Although Sakhipur town is well connected to Dhaka, the internal roads of the thana are not surfaced and become very difficult for walking or motor cycling after rain because of the sticky nature of the soil in the area. Sakhipur has many primary and high schools and two big colleges. The thana health complex is in the thana centre.

The staff structure of PROSHIKA at Sakhipur is as follows:

- Area Co-ordinator (1).
- Training Co-ordinator (3) (I/C Training in this book).
- Universal Education Worker (2) (Education Worker).
- Economic Development Worker (14) (Economic Worker).
- Development Education Worker (7) (Development Worker).
- Extension Worker (3) (Extension Worker).
- Accountant (1).
- Cashier (2).
- Caretaker (2).

An Area Co-ordinator is in charge of all the activities of PROSHIKA in his/her respective Area Development Centre (ADC). I/C Areas are supervised by the Zonal Co-ordinators (ZCs). For Sakhipur, the ZC is responsible for three ADCs and is based in Madhupur Thana. Accountant and cashiers are the only other office-based staff of PROSHIKA.

The major functions of I/C Trainers are to organise and supervise all training activities in their Area. In addition to organising the training programmes, I/C Trainers also follow up and monitor them. Universal Education Workers (Education Workers) are primarily responsible for establishing and running the Non-Formal Primary Education schools in their Area. They are assisted in their work by their Development Education Workers (Development Workers). Development Workers are mainly responsible for supervising the meetings of the PROSHIKA groups. They meet once a month in a village co-ordination meeting and all relevant issues are discussed in the groups. In Sakhipur there are 917 groups and the Area is divided into six units. There is one Development Worker for each unit, with around 19-20 villages. Economic Workers are responsible for the economic 'development' of the group members of PROSHIKA. For this, they organise skills training programmes, disburse credit and provide technical support wherever necessary. Some Economic Workers specialise in certain activities. In Sakhipur Economic Workers specialise in activities like apiculture, sanitation and housing, agriculture, irrigation, livestock, social forestry and sericulture. Usually, each Economic Worker has around 60 groups under his/her control.

PROSHIKA has a clear and formal system of monitoring its activities. Field workers are evaluated and monitored by the I/C Area, I/C Areas by the ZCs and the ZCs by central co-ordinators based in the Dhaka office. How then does the I/C Area monitor all his or her staff? The I/C Area of Sakhipur said that he uses a workplan for the month at the beginning of each month. But he (the I/C Area) does not disclose his own workplan. So, he suddenly goes to the places or meetings where field workers are supposed to be present. He also goes to the members and asks about their field workers, whether he/she meets them regularly, how does he/she behave with them etc. He also asks the children at the non-formal schools whether they know their field workers, how frequently they see him/her etc to test how effective and active he/she is.

PROSHIKA recruits its staff through open advertisements. Usually the educational qualification required for Development Workers are a BA, BSc or BCom degree. Most field workers join as Development Workers. In general, PROSHIKA groups are working well at Sakhipur. Some groups are very sincere about attending the training programmes. But many field workers informed me about ineffective groups. Splitting of the groups is also not uncommon, usually through internal problems.

Microcredit

A major activity of PROSHIKA is to provide credit. Most credit is repaid in monthly instalments. Most Development Workers interviewed complained that they are not given proper status; less than the Economic Workers, even though the Economic Workers look after the organisational matters of the groups. Since Economic Workers provide credit, group members give them all the status. Most Development Workers said that before sanctioning credit to any group, Economic Workers should discuss with them matters like the member's attendance at village co-ordination meetings or sending children to school. Development Workers think there should be more pre-conditions for sanctioning credit to groups.

Some long service field workers told me that in regard to microcredit, PROSHIKA is behaving like the street medicine sellers who draw public attention by singing, performing magic or playing with snakes in the markets, bus or train stations, river ports and terminals. After attracting the crowd, these businessmen start advertising their products. Like these businessmen, NGOs like PROSHIKA attracted clients through mobilisation and consciousness-raising and are now doing microcredit business (compare White, 1999). All long service field workers reported to me the changes that PROSHIKA has gone through. One field worker told me:

When I joined PROSHIKA 8 years ago, my work was to go to the farmers and to organise them in groups, motivating them through music and lectures. So, I learned how to sing and use the harmonium and the 'dhole' [drum]. People came to listen to our music and lectures. Some days, I found some people crying after listening to our music, which was describing how an affluent farmer became landless through the exploitation of moneylenders. My life was easy. One of my colleagues resigned from PROSHIKA to join CARE [an international NGO] to get a higher salary. One week later, he came back in tears saying that he misses us and does not like the overwork and higher salary. There was no credit programme but little skills or literacy training. Now I travel hundreds of miles, I have no leisure. I have to show a good repayment rate of my disbursed credit to save my job. To get money back sometimes I abuse my members. Now my life is full of tension. Many nights I cannot sleep due to the anxiety of what I shall do if I lose the job (Qamrul Islam, Training Coordinator, Sakhipur, February, 1998).

The RDRS (now RDRS Bangladesh)

Rangpur Dinajpur Rural Service (RDRS) is one of the oldest NGOs in Bangladesh, working in 29 thanas of six districts in northern Bangladesh. Serving around 200,000 households, RDRS employs about 1,500 staff.

Until 1997, RDRS was the Bangladesh field programme of the Geneva-based Lutheran World Federation's Department of World Service which operates development, relief and rehabilitation programmes in 25 countries. RDRS has, since the fieldwork, been transformed into a national NGO.

Rangpur-Dinajpur is one of the most poverty stricken areas of the country. An important aspect of the Bangladesh economy is that a major portion of its foreign currency comes from the remittances of those Bangladeshis working abroad, but in my RDRS study area I met no family with its close kin abroad. Northern Bangladesh is in general poorer than the rest of the country. It is the least industrialised area, with a predominantly agricultural economy. At the same time, most of the area where I worked is *char* land and the physical conditions and agricultural potential are not very different from Naria (my SCF study area). A major difference is that the men from the *char* areas of Naria go fishing in the rainy season and men from the Kurigram District migrate to Comilla or Mymensingh District for work during the sowing and harvesting seasons. After the recent inauguration of the bridge over Jamuna, Rangpur-Dinajpur seems likely to benefit from improved communications with the rest of the country, and isolation was a major problem.

My field work was conducted in three thanas of the Kurigram District - Kurigram Sadar, Ulipur and Rajarhat.

The presence of many NGOs in Kurigram is of interest. I could say that Kurigram is an NGO town. Walking or travelling by rickshaw, I counted around 20 NGO offices in the town. During the 1974 famine Kurigram was one of the worst affected areas, which attracted many NGOs. At the same time the activities of the NGOs and relief agencies seem to have made people in that area more relief-oriented than self-reliant. The recent decline in the supply of relief or NGO funds has made the work of the NGOs in this region more difficult because people think they deserve relief as they did three decades ago.

RDRS was born out of the 1971 war of independence in Bangladesh to provide relief to Bangladeshi refugees in the Cooch Behar district in India. RDRS later helped the returnees to resettle in the greater Rangpur-Dinajpur region and began an extensive programme of rehabilitation in that area. Since then RDRS has gone through three stages: Relief and Rehabilitation (1972-1975), Sectoral Development (1976-1987) and Comprehensive Development (1988 to present). Reflecting these changes, RDRS changed its name from Rangpur Dinajpur Rehabilitation to Rangpur Dinajpur Rural Service in the mid 1980s.

RDRS documents say that the 'development' philosophy of RDRS is essentially a holistic approach aimed to empower the poor. No doubt, these are good words. I agree that immediately after Independence there was a need for relief and rehabilitation work. This became more urgent when Bangladesh experienced a famine in 1974. But why did RDRS go for sectoral 'development' for 11 years and then abandon it for comprehensive 'development'? I would say, and shall later seek to prove, that the comprehensive 'development' programme is dominated by a microcredit programme which should in no way be termed 'comprehensive development'.

Table 3.1   The Stages of Development of RDRS

Year	Stage
1971	Lutheran World Federation sets up refugee camps
1972	RDRS relief and rehabilitation launched
1976	Sectoral programmes replace relief and rehabilitation
1978	Rangpur-Kurigram road completed
1979	First RDRS women staff to ride bicycles
1980	Haripur hospital handed over. Treadle pump passes its test
1983	First women's agricultural groups
1985	RDRS changes name to Rural Services
1986	Decision to reduce expatriate staff numbers
1988	Comprehensive Project absorbs sectoral system
1990	First Federations emerge
1993	First Comprehensive Group Graduations
1994	Merger of Comprehensive and Rural Works Projects
1995	Credit and bilateral projects take off
1997	Localisation of RDRS

Source: RDRS, 1996a.

Although the RDRS documents do not say so, Lutheran World Federation was facing a fund shortage and told RDRS. As long as RDRS remained a field programme of LWF, they could not take funds from other donors, so it was compelled to rename itself and change its status to that of a national NGO. This was done at the expense of budget cuts, going for bilateral programmes with other donors or NGOs (European Union, Helen Keller International, International Fund for Agricultural Development etc.) and above all job cuts ranging from 10 to 20 per cent, based on different estimates. One Thana Manager pointed out to me:

> *RDRS was born with a golden spoon in her mouth, now she has got a copper spoon so now she realises how difficult the world is.*

The Management Structure

The management culture of RDRS is bureaucratic. Not only the field workers but also mid-level and some senior managers complained about bureaucratisation over the last 27 years. Since becoming a national NGO, RDRS is building its head office at Rangpur on its own land. Earlier, all RDRS offices were on rented land and houses because foreign NGOs cannot buy any land assets in Bangladesh. The field level bureaucratic structure of RDRS is as follows:

Grade A: 1. Head of Field Programme (Rangpur based)
2. Heads of Co-ordination (Training and Extension) (Rangpur based)

Grade B: District Co-ordinators (based in their respective districts)

Grade C: Assistant District Co-ordinators (each District co-ordinator is assisted by one or two)

Grade D: Thana Managers, (based in each thana) and Sector Managers (based in two zones - east in Lalmonirhat and west in Thakurgaon)

Grade E: Extension Officers (based in all project offices)

Grade F: Assistant Thana Managers, mainly office based, in charge of two to three Grade G. Also Assistant Thana Managers (Credit). Both based in thanas.

Grade G: Union Organisers, the RDRS field workers (one or two in each union)

Grade H: Night guards and similar staff.

The structure above clearly shows the bureaucratic nature of field-level management in RDRS. All recruitment in RDRS is through advertisements in the national newspapers and internal notices.

Facilities and Benefits Enjoyed by Field Level Staff

In addition to their basic monthly salaries, all staff of RDRS get a house-rent allowance of 65 per cent of their basic salary, a medical allowance and a festival allowance once a year. Other benefits, like the yearly increment, provident fund and gratuity, are also available to all staff of RDRS. In addition to paid maternity leave (180 days) women staff get a children's allowance for a year for up to two children.

All field staff get vehicles for their field work. All Thana Managers and Assistants get motorcycles on a hire-purchase basis.

All payment for staff vehicles is by deductions from monthly salaries. For fuel, all staff entitled to a motorcycle get Tk 1.25/km, the ceiling for a Thana Manager being 900 km/month and for others 700 km/month. Most Managers interviewed complained that the ceiling is inadequate. All Union Organisers get around Tk 100/month for bicycle maintenance while all Managers get around Tk 100/month for maintenance of their motorcycles.

Table 3.2   The Hire-purchase System of Vehicles in RDRS
Bangladesh

Vehicle	Payment System
Motorcycle - Honda, price Tk 67,000 (for Managers except Credit)	Tk 52,000 to be repaid in 4 years and staff can get a new motorcycle after 5 years
Motorcycle - Hero Honda, price Tk 60,000 (for Managers except Credit)	Tk 45,000 to be repaid in 4 years and staff can buy a new motorcycle after 4 years
Bicycle - Tk 4000 (for all Union Organisers	Tk 3200 to be repaid in two years

Source: RDRS Bangladesh official documents.

Microcredit

Every morning, the Union Organisers leave their home or mess to see their clients and collect or disburse money. Although RDRS provides several services to its clients (mentioned earlier), all field workers are preoccupied with credit. Most field workers interviewed complained that they cannot give enough time or importance to other services due to the credit system. Compared to the SCF 'partner' NGOs, RDRS field workers have to work longer hours, even at night, particularly the men.

Although RDRS field workers have to provide many services to their clients, most of their time is spent on credit. Although they must talk about the education programme, plantation work, fishery and livestock projects etc. their job performance is measured mainly on the repayment rates of the credit they disburse. I found most to be very hard working, but some were poor at keeping records, which caused many misunderstandings with their clients.

The SCF 'Partner' NGOs

This case study is of very small local NGOs working with a large international NGO since 1996 in Naria Thana of the Shariatpur District. SCF (The Save the Children Fund UK) is one of the leading international NGOs in Bangladesh. It started its work in Bangladesh soon after independence in 1971. One major activity of SCF (UK) was to enable its beneficiaries to cope with flood, which is a major disaster in Bangladesh. SCF (UK) has Coping With Flood programmes in some flood-prone districts of Bangladesh. In 1996, SCF (UK) decided to withdraw from running its own local projects in the study area, and work with local 'partner' NGOs. Their ex-field workers set up local NGOs to become 'partners'.

The Study Area: Naria Thana

Shariatpur, my research area, is located to the south-west of the capital Dhaka. Although Shariatpur is physically only 35 miles away from Dhaka, it has a poor transport infrastructure network. Most of the roads in the district are in a dilapidated condition and the situation deteriorates during the rainy season. It is better to travel to Shariatpur from Dhaka by boat, which takes around 5 hours during the rainy season. This form of travel takes longer and becomes risky during the dry season due to the drastic fall in the water-level in the Padma (Ganges) which exposes *chars* or temporary islands in its course. Shariatpur does not have a railway network.

My interviews took place with the SCF 'partner' NGOs in Naria Thana of the Shariatpur District. All the three where these 'partners' work (of which two are *chars*) have primary (Year 5) schools and high schools (Year 10) but no colleges. Most students go to Naria College (at the thana headquarters) or Sureshwar College (at Sureshwar Union), of which Naria is a government college. The only health facility available in Naria is at Mulfatganz. Life on the *chars* is really difficult, with no electricity. There is electricity at Gharishar, but power cuts are a regular phenomenon and the voltage is so low that it is useless.

There are some surfaced roads in Gharisar but they are simply absent on the *chars*. The major crops are jute, paddy, potato and chilli. Many people leave for the coastal areas of Bangladesh at the beginning of the rainy season and come back at the beginning of the winter. While I was in Naria many fishermen (seasonal) were coming back with money after fishing. Many people from my study area are working abroad, mainly in the Ukraine, Italy and Malaysia. A major problem in Naria is river erosion. The earlier SCF (UK) office was totally washed away by the Padma about two years ago. The present office is around 200 yards away from the river, but could be washed away again.

Direct Work of SCF (UK)

The history of the direct work of SCF (UK), up to 1996, is important for an understanding of Naria Thana. Shariatpur (particularly Naria Thana) experienced severe flooding in 1987 and again in 1988. SCF (UK) provided relief to the flood-affected people during both floods. Later, SCF

(UK) was still a direct action NGO. It decided to continue its operations in Naria Thana and started its credit and savings project in 1992. The activities of SCF (UK) then had several components - credit and savings programmes for its clients, Traditional Birth Attendant (TBA) training, sinking of tubewells, training village doctors (without formal degree), etc. The credit and savings programme was and is aimed at alleviating the poverty of its clients. SCF (UK) clients formed 5 to 10 member groups. All of the members were women and the field workers of SCF (UK) were also women. SCF (UK) advertised for field workers for its programme and both men and women from Naria Thana applied. After conducting the interviews, the Dhaka office directed the Naria office to recruit only women. This created some discontent among the men applicants, and they protested against this move by SCF (UK). The SCF (UK) Naria staff approached local leaders and brought the situation under control (information given by field workers of 'partner' NGOs, and by SCF (UK) staff at Naria office, November, 1997).

The birth project was mainly directed to train traditional birth attendants to ensure safe childbirth and provide neo-natal and ante-natal care. SCF (UK) trained 262 by 1997. The main objective of sinking tubewells (mainly shallow) was not only to create awareness about using safe water but also to ensure regular supplies of potable water among the clients and their neighbours. In order to maintain these tubewells, SCF (UK) trained two women clients for each tubewell. By early 1997 SCF (UK) had sunk 211 tubewells. The training of the village doctors was directed to provide better health-care facilities to people in the villages who cannot afford to go to the nearby health-care centres. In other words, the project was aimed at training the village doctors so that they could provide better health-care advice to their patients.

For the credit and savings programme SCF (UK) had three centres in each union, with one field supervisor and a few supporting staff. SCF (UK) had on an average around 30 field workers in Naria Thana throughout its period of direct operations. SCF (UK) continued its credit and savings programme up to June 1996. In July 1996 SCF (UK) handed over its credit and savings programme to the three NGOs formed by its former field workers. For a discussion on the pre-handover activities of SCF (UK) in Shariatpur, see Edwards (1999).

The Change to 'Partners'

According to the then Country Director of SCF (UK) the reason for handing over their activities to their 'partners' were:

1. In mid 1996, SCF (UK) field workers were getting a salary of about Tk 4000, which was a handsome sum compared to the qualifications and salaries of field workers of other NGOs and Government Organisations (GOs). SCF (UK) used to pay its field workers regularly with one festival bonus and a gratuity each year. The cost of operating the credit and savings programme reached a level which seemed very high compared to other NGOs in neighbouring villages (see Malhotra, 1997).

2. SCF (UK) decided that it is worthwhile to help the growth of local organisations who would be better able to do the 'development' work by themselves. SCF (UK) also decided to provide the necessary technical advice to the new NGOs for a couple of years during the transition period.

3. The policy makers at the SCF (UK) head office realised that the organisation had become too bureaucratic in Bangladesh and decided to reduce costs. During the mid-nineties, SCF (UK) retrenched many of its staff (management, field and ancillary) (interview with Simon Mollison, October, 1997).

When SCF (UK) decided to hand over its activities to 'partner' NGOs, that period was a critical time for the SCF (UK) management, field workers and clients. After prolonged tension, the field workers of SCF (UK) in Naria decided to form three new NGOs in their respective unions. SCF (UK) is now trying to hand over its activities to other NGOs in another district, Jamalpur.

An interesting aspect of this handover is that the people in SCF (UK) refer to it in different ways. The Country Director, Simon Mollison, does not like to accept the NGOs newly formed by the former SCF (UK) staff as 'partners', but thinks of them as independent NGOs (interview with Simon Mollison, October, 1997). Formally they are independent NGOs, but other senior staff of SCF (UK) in Dhaka think of them as partners. In many cases, however, SCF (UK) behaves like a donor, and a degree of dependency on SCF (UK) for funds and technical advice is clearly visible.

The three NGOs have a staff structure which is rather flat by Bangladeshi standards because of the early stages and small scale of their activities (Table 3.3). All recruitment of field workers in SCF 'partner' NGOs are through advertisements in Union Council offices, and notice boards of the local schools. Some clients canvass their field workers to recruit their daughters who have 8-10 years schooling.

Table 3.3   The Staff Structure of the SCF 'Partners' at Naria Thana, Shariatpur District

NGOs	Director	Accountant	Field Workers	Total
CWDS (Char Atra Union)	1	1	4	6
Upoma (Gharisar Union)*	1	1	8	10
Sakaley Kori (Naopara Union)	1	1	4	6

*Upoma has one *Aya* (domestic worker) for office services such as cleaning, serving refreshments and other ancillary work.

Source: The NGOs and field work.

Why did the field workers decide to form their own NGOs and continue their activities? Field workers identified several reasons to me:

1. While the field workers were working for SCF (UK) (1992-96) they disbursed a large amount of money (more than two million Taka) which would have been very difficult to recover. This could have been very embarrassing for SCF (UK) and for the field workers who were members of that community which could also default. Field workers told me that SCF management staff could leave because they were 'outsiders', but they themselves could not. Field workers felt that it would be very humiliating for them and their families if they stopped giving credit to their clients. This would be a double disgrace for their families, since all field workers were women with an average education of about 8-10 years schooling.
    Field workers told me that the benefits and services which they received before handover brought enormous social and economic advantages to them. Most saved from their monthly salary. Some bought ornaments, helped their husbands or brothers in going abroad, starting businesses or getting jobs (through bribery, which is very common in Bangladesh). So, through working as a field worker for SCF (UK) these women got some status which they did not like to lose by giving up working. So, they formed their own NGO.
2. SCF (UK) agreed to give 'partners' the Loan Revolving Fund (LRF) and some furniture. SCF (UK) staff also helped them in getting registration as an NGO from the government, and signed a three year agreement with the new NGOs outlining the modes of 'partnership'.
3. One major reason for the former SCF (UK) field workers to start their own NGO was that they wanted to work independently. Initially, they

had some confusion and misunderstandings among themselves and with the management of SCF (UK). Most of the field workers said that they enjoyed the independence of managing their own organisations. Previously, they were simply the implementers in the highly bureaucratised SCF (UK) structure.

In addition to funds, SCF (UK) provides the field workers of partner NGOs with services like training for capacity building, office management or sustainable group development. It also provides help to new NGO staff in visiting other NGOs to see and learn from their activities.

Most of the present field workers and directors of the new NGOS (who were formerly field workers) have experienced a steep fall in their salary. Field workers get around Tk 1500-1800. per month, directors and accountants roughly Tk 200-300 more. All also get a festival allowance and a gratuity every year, each equivalent to one month's salary. Most staff in the new NGOs are unhappy at this huge fall in their salaries. Some of them blame the SCF (UK) management for spending too much money at the management level, which unrealistically raised the costs of operating the credit and savings programme. The Country Director, who played a key role in re-organising the activities of SCF (UK) and handing over the credit and savings programmes to the new NGOs, told me that field workers are now getting the salary that they deserve. He blamed the previous SCF (UK) administration in Dhaka for the earlier high administrative costs. Now the money for their salaries and office maintenance costs comes from the interest of their disbursed credit and interest from the savings of their clients (interview with Simon Mollison, October, 1997). Most field workers complained to me that they now have to bear the brunt of lower expenditure from their lower income. Some said they cannot meet many basic demands for their families and near and dear ones, which has really aggrieved them.

Although SCF (UK) has an office with skeleton staff at Naria in Shariatpur, it is gradually winding up its operations. The new NGOs still need technical advice, suggestions for management decisions and above all for the proper utilisation of the SCF (UK) funds that they have as a revolving fund. This is important since many clients stopped paying their instalments during the hand over due to fear of being cheated by a 'Christian' foreign organisation. If there is any weakness among these new NGOs, other NGOs in the area will try to create division and mislead the clients. SDS (Shariatpur Development Society, a local NGO) tried to do so during the hand over. I found that SDS was trying to lure away members of the SCF 'partner' NGOs, and is always trying to propagate a bad name against them. When I was in Naria, the SDS Director came to see me and tried to do the same.

A major feature of the new NGOs is that their management structure is flat and staff usually have an amicable relationship between themselves. On close inspection, it seems that they have to remain united for the very survival of their NGOs. If someone goes on leave, other field workers share her work. Things are not so good in all cases. During the formation of Upoma, a clear rift erupted regarding who would be the director and the accountant, which had to be solved through a secret ballot under the supervision of SCF (UK) staff (information from field workers of Upoma and SCF (UK) managers at the Naria office, November, 1997).

'Partnership'

The way SCF (UK) was working before handing over and the way the SCF 'partner' NGOs are working now seems to be very different to the (interim) SCF (UK) Bangladesh Country Strategy Paper (1997) which says:

> *Nearly all development work in Bangladesh is targeted at the great mass of 'poor people' but actually neglects the poorest and most marginalised within this group. Thus, for example, the famous Bangladesh credit programmes typically do not benefit the poorest strata of society and, although they are targeted at women, may tend to exploit or burden women further. At the same time, the poverty alleviation programmes which much of the development community in Bangladesh are engaged in treats poverty as an almost purely economic phenomenon. Many of the most serious situations in Bangladesh are not the 'inevitable consequences of poverty' that both NGOs and policy makers seem to see them as. Such situations are associated with poverty but are caused by social factors. They are not likely to be solved within the next fifty years by the slowly rolling poverty alleviation programmes which dominate the development scene in Bangladesh. ...Many of the problems of the most poor and marginalised are both economic and social in nature but social problems are especially neglected in much development practice. We will therefore focus much of our attention on them* (SCF UK, p. 3, 1997).

The activities of the 'partner' NGOs of SCF (UK) do not reflect this statement in the Country Strategy Paper, and the relationship between SCF (UK) with its 'partners' is poorly defined. But the term 'partnership' seems rather misleading to me because the partnership deal with the partners is titled as 'Agreement for Management of the SCF (UK) Revolving Fund' with its partners which says:

1. The loan programme should be kept confined to certain Unions (Gharisar in the case of Upoma).
2. Section 2 says SCF (UK) has agreed to allow the 'partner' to use the RLF for three years but its ownership will remain in the hands of the SCF (UK).
3. Section 4 says 'partners' are bound to show all the books, registers to SCF (UK) at any time.
4. Section 6 says 'partners' can use the money from the service charge for maintaining their office or the staff of RLF but the 'partners' are bound to obey any order from SCF (UK) in this regard.

This seems less like 'partnership' than donorship between the SCF (UK) and the NGOs.

Microcredit

Usually the new NGOs (Upoma, Sakaley Kori, Char Atra Mahila Unyayan Samity) provide credit of up to Tk 3000. The clients have to repay the loans in weekly instalments over 50 weeks, and to save around Tk 5-10. per week to get a bonus of around two per cent every six months. Credit taken by clients of the new NGOs is usually used partly to pay for the costs of farming, fishing, cattle raising etc. Clients also use credit to buy or build new houses, pay examination fees for their children for appearing at public examinations and to pay the costs of the marriage of their daughters. Most clients wanted an increase in the amount of credit to at least Tk 5000, which seems impossible for these new NGOs, given their financial condition.

Summary and Conclusion

Before I end this Chapter, I want to summarise the activities and benefits enjoyed by the field workers of my study NGOs. Some NGOs in Bangladesh have experienced tremendous growth (both in terms of activities, funds, staff and clients), mainly due to the availability of funds and donor preference. Undoubtedly, providing credit has become the major activity of NGOs in Bangladesh, except MCC (Mennonite Central Committee). Credit provision has made the activities of NGOs more businesslike. The other aspects of 'development' are of less importance to all the NGOs except MCC. This nonconformity is possible for MCC due to its comparatively steady flow of funds, and given the highly missionary nature of that NGO. So, MCC can afford to avoid the recent fashion of microcredit. However, it should be noted here that MCC has also gone into 'partnership' due to fund shortages. This Chapter also shows that NGOs in Bangladesh are not Grass Roots Organisations (GROs).

Table 3.4   The Basic Information about the Study NGOs

MCC Bangladesh (international, missionary)

Activities:	Agricultural support services (including training), Research, Poultry, fishery and livestock development, Savings and capital generation, Health education, Employment generation
Staff:	141 (both men and women)
Working Area:	31 thanas, 14 districts
Year of Establishment:	1970

PROSHIKA (large national)

Activities:	Credit, Savings, Education, Employment and income generating activities, Environment, Health, Women's development', Housing, Disaster mitigation, Advocacy.
Staff:	2988 (both men and women)
Working Area:	156 thanas, 45 districts
Year of Establishment:	1976

RDRS Bangladesh (regional, national)

Activities:	Credit, Savings, Legal Aid, Agricultural support services, Education, Health, Income generating activities like poultry, fishery, livestock etc.
Staff:	1500 (both men and women)
Working Area:	29 thanas, 6 districts.
Year of Establishment:	1971

SCF 'Partner' NGOs (local)

Activities:	Credit and savings
Staff:	22 (all women)
Working Area:	3 unions, 1 thana, 1 district
Year of Establishment:	1996

*Note: Bangladesh has 490 thanas and 64 districts (BBS, 1997). All staff are local full-time staff.

Source: The NGOs.

If you look at the benefits enjoyed by the field workers of my study NGOs, it is clear that the best package for women is enjoyed by the MCC staff (see Table 3.5). Here again the international management and missionary values seem to me the major reason for these better benefits (PROSHIKA's package for men is also good). Overall, the disparity in benefits enjoyed by different types of NGOs in Bangladesh is appalling. SCF 'partner' NGOs have very few benefits, formal or informal.

Table 3.5   Benefits Enjoyed by the Field Workers of The Study NGOs

MCC Bangladesh	PROSHIKA	RDRS Bangladesh	SCF (UK) 'Partners'
Salary (4810)	Salary (5016)	Salary (3838)	Salary (1427)
House rent (50% of basic)	House-rent (65% of basic at thana)	House-rent (65% of basic)	Festival allowance (1/year)
Medical allowance	Medical allowance (general and special)	Medical allowance	Gratuity
Motorycle/ bicycle	Provident fund	Festival allowance (1/Year)	Leave (casual 30 days, maternity 90 days)
Fuel, Gratuity	House building loan	Increment	
Festival bonus (1/Year)	Gratuity	Provident fund	
Umbrella	Maternity leave (84 days)	Gratuity	
Raincoat	Motorycle/ bicycle	Maternity leave (180 days)	
Increment	Maintenance	Child Allowance (1 year/child)	
Maternity leave (90 days)	Fuel	Motorcycle/ bicycle (hire purchase)	
75% cost of child delivery	Increment	Fuel and maintenance	
Child Care (maximum 2 Years including aya or nanny)	Festival bonus (1/year)	Food allowance	
		Leave	

*Note: Salary is the gross salary that field workers got during the last month of the survey which includes basic salary, house-rent and medical allowance (if available). All figures are in taka.

Source: Field survey and NGO documents.

In this Chapter, I have described the NGOs with whom I worked for my research. I have also described the geographical and socio-economic environments of the areas where I worked. In the next Chapter I shall elaborate on the socio-economic background of my research subjects - the field workers.

Chapter 4

The Careers of Field Workers

This Chapter deals with the socio-economic background of the field workers of NGOs in Bangladesh. I believe my research will help the NGOs in the South to utilise their field workers better. To do that it is essential to know who the field workers are and what are their careers and previous personal and professional problems. In this Chapter I shall try to answer these questions. Who are the NGO field workers? Why and how have they become the field workers for these NGOs?

I have set out to establish the social backgrounds, and educational status of field workers in selected NGOs. This social structure largely determines that they are from a specific stratum of rural society. Education makes a difference to the career and life of a man or woman in Bangladesh. The education system in Bangladesh is divided according to social class. Due to their families' low socio-economic position, field workers receive a general or Bengali education. To get good results they would have needed private tuition, which they could not afford, and due to their poverty they may have had to drop out of school at any stage. I have found evidence of that, which I shall describe in this Chapter.

The NGO field workers interviewed here are from middle-class families in rural Bangladesh. Ninety-one per cent of the respondents to the questionnaire survey were born and brought up in villages (unlike Garain, 1993). This status defines their level of education. I used to believe that education could make all the difference in their lives. My belief was shattered when I met the field workers. It is difficult, I should say impossible, to get a good job (in Bangladesh, a government job) when you have neither achieved a good degree nor have family or social links and money for bribes. All the field workers would have preferred government jobs, but failed to get one (White, 1999); I shall elaborate this below.

When I describe the backgrounds of the field workers it will become clear how weak their positions are, both in their community and in their organisations. An interesting exception is that most field workers in the SCF 'partner' NGOs are from influential families; I shall show how the influence of the families of these women was exploited by their NGOs to get access to the community and to get borrowed money back on time. In this Chapter I shall also describe the backgrounds of the immediate superiors of the field workers and explore the differences between the background of current field workers and of former field workers who have managed to climb a few steps up the long ladders of their organisations.

The field workers of NGOs in Bangladesh are not only from a specific social stratum, they are also from a specific age-group. Field work requires demanding physical exercise, so field workers are from the most youthful age-group (see Table 4.2).

Family Backgrounds of Contrasting Field Workers

The fathers of most field workers work or worked in farming, petty trading or ancillary jobs in government or semi-government offices. There is an important division here. Most field workers are posted away from their homes. In contrast, as we have seen most SCF 'partner' NGO field workers are from influential local families. These are the NGOs where field workers can work in their own village and all field workers are women. Above all, these NGOs are particularly local and small. I talked to the mid-level and senior managers of SCF (UK) and all of them preferred to allow their field workers to work in their own village. Why? There are several reasons:

SCF (UK) (like MCC, Mennonite Central Committee) is regarded as a Christian, foreign NGO. A major problem of access for such NGOs is that they are suspected of being involved in Christian evangelism. SCF (UK) therefore needs access to local leaders and influential people. A major way of doing that is to recruit from those families. The daughters of rich families in rural Bangladesh will not work as field workers for NGOs, for this would be a matter of disgrace, which leaves the daughters of influential middle-class families. This policy helped SCF (UK) get into villages and to get their work done easily, because their target groups were poor families who felt obliged to repay money borrowed from the daughter of an influential family. Also, when the daughter of an influential family travels about, she is not usually teased by the local youth, because they know the social strength of this girl's father. In that sense, the women field workers of other NGOs are rather unlucky because they have to ride bicycles or motorcycles and experience teasing, at least at the beginning of their careers.

Another advantage of recruiting local girls is that clients can trust the field worker and her NGO and can safely give her money as savings. Most NGOs start their work with their clients by forming groups and saving money in the group fund. Why should the client save money with a foreign 'Christian' NGO? The answer from a field worker of an SCF 'partner'

NGO is, 'I am from your village, you know my family, I cannot run away with your money, can I?' This solves the major problem of group formation and setting up a microcredit programme.

Employing local youth (women in the case of SCF 'partners') also has its disadvantages. When SCF (UK) was gradually winding up its operations in Naria it retrenched many field workers in stages. One field worker who was dismissed was the daughter of a local *matbar* (leader). Her brothers went to the SCF (UK) managers at the Naria office and threatened not only to tell local people not to repay, but also to attack the SCF (UK) Naria office, again with the help of the local people. The SCF (UK) manager then cancelled the dismissal and brought the situation under control. Mainly because of this risk, other NGOs (MCC, PROSHIKA, RDRS) do not allow their field workers to work in their own villages. Even so, why did RDRS (Rangpur Dinajpur Rural Service) and MCC (Mennonite Central Committee) not try to get access to their prospective clients through the local influential families like SCF (UK) did? Although RDRS and MCC started work in relief, and as foreign NGOs (MCC is still a foreign NGO), and switched over to 'development', I asked them why they did not work like SCF (UK)? All senior and mid-level managers of MCC and RDRS told me that their organisations anticipated problems which may arise from recruiting local youth. All NGOs avoid conflict with local leaders and keep links to the local power structure. As a policy MCC, PROSHIKA and RDRS do not allow their field workers to work in their own villages. Almost all field workers, and all mid-level and senior managers of these three NGOs interviewed (see also Fowler, 1997), supported this policy.

Educational Backgrounds of Field Workers

Field work for NGOs is not an attractive job to well-educated youth. Most mid-level and senior managers of NGOs complained to me that field workers are very poor at record-keeping (I found the worst were in RDRS). When I was working at Sakhipur (with PROSHIKA), I found one woman field worker who had to write her application for leave four times. The Area Coordinator complained to me about the level of education of that field worker. When my field assistants interviewed the field workers, some 90 per cent gave the wrong definition of their target groups, as they did when talking to me. All NGOs in Bangladesh have a definition of their target groups. Soon after joining, whether through training or by instruction from their immediate superiors (field level managers), field workers learn these targets. During the survey, they were asked about their salary and other benefits. Hardly any field worker knew how much of their total salary was the basic salary and how much was the other benefits (house rent, medical allowance, transport allowance etc.). Legally they should, but many have not asked.

So, why are field workers so poor at reporting basic facts? A major reason must be poor training.

Table 4.1   Educational Background of the Field Workers

Academic Qualification	Name of NGOs								
	MCC Bangladesh		PROSHIKA		RDRS		SCF 'Partner NGOs		**Total**	
		%		%		%		%		%
Primary	1	2	0		0		6	40	7	7
SSC	17	42	0		7	20	8	53	32	30
HSC	10	25	0		13	37	1	7	24	22
Dip Eng.	0		1	6	0		0		1	1
Graduate	13	32	8	50	13	37	0		34	32
Postgraduate	0		7	44	2	6	0		9	8
Total 100%*	41		16		35		15		107	

Note: Primary means 5 years schooling, SSC means 10 years, HSC means 12 years, graduate means a BA/BSc/BCom/BSc (Agr) degree holder, postgraduate means a MA/MSc/MCom/MSc (Agr). *Percentages are rounded so they may not add up to 100 per cent.

Source: Field Survey.

Table 4.1 shows that, except in PROSHIKA, most field workers have SSC or HSC qualifications.[1] There are several reasons for this. One major one is that as a policy SCF (UK) and MCC (Mennonite Central Committee) do not welcome men or women with degrees as field workers. More specifically, they think field workers do not need to be highly educated. RDRS (Rangpur Dinajpur Rural Service) has recently started to recruit field workers with degrees but most of the previous field workers have SSC or HSC qualifications. PROSHIKA is one of the largest NGOs in Bangladesh. Jobs in PROSHIKA are more secure than in most other NGOs. Staff turnover is still very low in PROSHIKA and due to its reputation it can attract graduates and postgraduates. I found many agricultural postgraduates working for PROSHIKA and RDRS. All told me that they have joined these NGOs because there were no other jobs available. New

---

[1] SSC is 10 years schooling and HSC is 12 years schooling.

field workers in RDRS and PROSHIKA are at least graduates. When NGOs like RDRS and PROSHIKA advertise for field workers they require graduates, but many with master degrees apply because unemployment among educated youth is very high in Bangladesh.

So, why is it the policy of some NGOs to recruit field workers with only an HSC when the standard of education at this level is very poor, and is poorest in the rural areas? Most field workers told me that less educated staff (especially women field workers) do not argue with their superiors. Less educated field workers remain more obedient. Another complaint the mid or senior level mangers made to me is that more educated field workers always try to get government jobs so they are not serious about their NGO job until they reach 30, the maximum age to get a government job in Bangladesh (compare White, 1999). All the immediate superiors or field managers resented the policy of recruiting people with MAs to their NGOs. (I shall discuss the preferences of the field managers for field workers below). Because family income is low, students have to look for jobs. If they get one they simply drop out of education. In the case of women, marriage is a major reason for dropping out, followed by distance of college from home and then high transport costs. One RDRS (Rangpur Dinajpur Rural Service) Assistant Thana Manager told me that he had to support his family when his father became sick and could not work. Most field workers told me that they did not feel motivated to pursue higher education when they know they will not get good jobs, even with a higher degree, due to their poor family connections. For women the general belief is that higher education will have little value. The older they become the less will be their prospects of a good marriage, which works as a major disincentive.

Age Structure of the Field Workers

This topic might appear tangential to the topic of this book, but I shall demonstrate its relevance. Daniels (1988) found that nonprofit organisations in the US look for committed recruits - those who enter at an early age, get the approved training, identify with the collective activities of their occupation, and leave only to retire. This is the same in Bangladesh, except for the last. In my sample (Table 4.2), more than 84 per cent of the field workers were under 36 years of age and 65 per cent were under 31. Very few had hopes of promotion or of staying in their post after 40. Field workers of NGOs are at the most productive stage of their lives. Field work requires demanding physical activity and good health, but very few can retain that degree of physical fitness for more than 20 years in the South. I suspect this is less than 20 years for women, considering the nutritional status of women in middle-class rural households in Bangladesh. So, what will happen to these thousands of field workers when they reach 40 or 45? I shall discuss this below.

Field workers told me that this is an exploitation of young workers by many NGOs. NGO senior managers claim that it is due to poor performance and poor education that most field workers do not get promoted. Among the four NGOs I studied, I found PROSHIKA field workers to have the best promotion prospects, although most still complained of corruption in the system. Most field workers told me that by the time they are 45 their salaries reach a point where the NGO can recruit two or three new field workers for the money. One way out for the field workers is to leave the job. Otherwise, as their health deteriorates and, due to lack of promotion, their morale declines, their performance will be poor and they will face unwelcome transfers and, at worst, redundancy.

Table 4.2   Age Structure of the NGO Field Workers

Age								Name of NGOs				
	MCC Bangladesh		PROSHIKA		RDRS Bangladesh		SCF 'Partner' NGOs		**Total**	
		%		%		%		%		%
<25	6	14	1	6	6	17	7	47	20	18
26-30	21	50	6	48	17	47	7	47	51	47
31-35	9	21	7	44	5	14	1	7	22	20
36-40	2	5	1	6	3	8	0		6	5
41-45	2	5	1	6	5	14	0		8	7
46-50	2	5	0		0		0		2	2
Total	42		16		36		15		109	
100%*										

*Note: Figures are rounded so they may not add up to sum as 100 per cent

Source: Field Survey

If NGOs exploit the youth of their field workers, how can they save their clients from exploitation by society and the market? If this is the future which field workers face, how can these change agents bring changes to their clients' lives to make a better future?

Future Plans of the Field Workers

Given this problem of job insecurity, the fear of ageing always haunts the field workers. All told me that after 40 or 45 it will be difficult for them to keep moving from place to place and work as hard as their organisations demand. The future for many unmarried women field workers is even more gloomy. Most told me that they do not know when they will marry; only if their future husbands permit will they continue to work or try for other jobs (see below).

Most field workers have plans. Most women save Tk 200-300 a month in the pension schemes. Some have ornaments specially made, an intelligent saving considering the value and prestige these bring. In Bangladesh, the best investments against inflation are land and ornaments. Banks are not popular in rural areas and interest rates are low, also services are poor. Most men want to be self-employed because they will not find any other job. Field workers told me that after working for NGOs for several years they have achieved a good idea of how to manage livestock, poultry or fishery projects and small businesses. One RDRS (Rangpur Dinajpur Rural Service) field worker told me that he is planning to open a computer training centre with the money from his gratuity and in the meantime his younger brothers (whom he is helping financially) will finish their education. One women field worker of an SCF 'partner' NGO told me that she has bought a cow and if she can continue her job for two more years she will buy another, and these will give her a regular income.

Many field workers (particularly the younger ones) told me that they will take MSS (Masters in Social Sciences) or MA examinations. This could help them to join private firms or private schools where, unlike with government jobs, there is no age bar. If they need money for bribes they will use their gratuities. One PROSHIKA field worker told me that he will take a law degree and then enter the legal profession.

Table 4.3   The Life Cycle of NGO Field Workers

Age	0-30	31-45	>45
Activities	• Education: SSC, HSC, BA/BSc/BCom, MA/MSc/ MCom/MSc (Agr) • Search for jobs in government, schools, NGOs. • Marriage (all women, most men)	• Working in the NGOs • Search for jobs in private schools. • Search for jobs in better NGOs • Marriage (all men)	• Promoted to mid-level or senior level management of NGOs (very few) • Redundancy/ retirement from NGOs • Self-employment (most men) • Business, work in private schools or firms, agriculture (all remaining men) • Housewife (all women)

How They Became Field Workers of Their Present NGOs

It does not surprise me that all field workers' first choice of job is in the government, as I myself started my career as a civil servant. Soon after finishing my MSc. in Geography at Dhaka University I was offered a job in the research section of PROSHIKA, which is now the IDPAA (Institute for Development Policy Advocacy and Analysis). PROSHIKA also told me that if I worked satisfactorily for them for three years they would send me to the UK to do a PhD at Bath. When I told my father (a former civil servant) about it he rejected it outright and said, 'Is that a job!' He advised me to sit the civil service entrance examination. When I left the civil service, many of my friends, relatives and colleagues in both civil service and university jobs expressed their disapproval. Government service in Bangladesh is almost a job for life; it gives immense power and status and above all a handsome pension and gratuity on retirement. Due to rampant corruption, many government officials make a great deal of money. Compared to government jobs, jobs in the NGOs are insecure, have lower salaries and benefits, and the staff are burdened with heavy workloads, low status, little power and no pension (Siddiqui, 1987). Save for one field worker in RDRS (Rangpur Dinajpur Rural Service), none of my respondents had ever worked in a Governmental Organisation. He had worked for the state-owned Bangladesh Railway, which he left because of the golden handshake programme. Bangladesh Railway is one of the largest loss-making state concerns in Bangladesh. Under a voluntary termination programme, funded by the Asian Development Bank and The World Bank, many workers left for handsome financial benefits.

My findings are similar to those of Woolcock (1998) on the Grameen Bank. He found that most field workers join it after failing to get a job in government (compare White, 1999). A major reason for choosing it was its reputation, but they told Woolcock that they would 'leave tomorrow' if they got a better paying job. Some field workers (all men) had worked in business or in private schools. Their major reasons for leaving these jobs were an irregular or poor salary or loss of the jobs; all these jobs are very insecure.

Once again, why do these people join NGOs as field workers? One RDRS field worker gave me the reason:

When I passed my BA I applied for all the jobs advertized in government that I could. My father is a farmer. He could not afford to give me Tk 80,000 to go to Middle-East or Tk 120,000 to go to Japan. One thing I could do is to marry a girl whose father could give me the money to go abroad. When I learned that RDRS were recruiting field workers, I applied. I had to deposit Tk 10,000. I borrowed it; nobody objected because RDRS is a well known NGO in my area. People see RDRS people on motorcycles and jeeps and they have been working in our area for twenty years (Abdul Quddus, Rajarhat, December, 1997).

Despite the poor salary, many field workers work in the NGOs not just because of high unemployment but for social reasons. For someone with a BA or MA, employment in such a service is much more prestigious than farming or managing a poultry farm, grocery or business, even if they had the money or intention of doing so. As a RDRS field worker at Ulipur told me:

The hard work that I do, leaving my family far away, I could earn much more by managing a poultry farm than I get from my NGO. But I cannot do that. My murrabbis (senior relatives) will say what are you doing with a BA degree? You are doing farming! It is the work of the illiterate. I work for an NGO, it has an office, I ride a motor cycle, this is a much better identity than saying that I am a farmer or a small businessman. This will help me in arranging my daughter's marriage because the bride's father is a job holder. What a society we are living in (Ershad Islam, December, 1997).

In a society where entrepreneurship is little respected, this is only to expected.

Case Study: The Story of Mr Mokhlesur Rahman (Maijdee, March, 1998)

The story of an MCC (Mennonite Central Committee) field worker shows how a middle-class rural Bangladeshi youth became a field worker for an international NGO. His story indicates the history of multiple jobs, which is common to many men field workers, and the frequent impact of events on their personal lives.

Mokhlesur hails from Devidwar Thana of the Brahmin Baria District. He passed the SSC in 1983 and took his HSC in 1985, but could not pass. His father was a tailor and involved in petty trading. His father died in 1983 just before his SSC examination. He has one brother and two sisters. He is the eldest son of his father; his eldest sister has five years schooling and is married to a farmer. His second sister had two years schooling and her husband is a farmer and wood trader. His younger brother is studying, in class ten.

After the death of his father, his family was drawn into deep financial crisis. He started to look for a job but without success until 1986. In 1986 he got a job as a salesman in a hardware store in Dhaka, where he worked for one year. In that job his salary was only Tk 150 per month, with free food and accommodation. With this small amount of money he could barely maintain himself. He tried to get a government job without success.

He came back to his village in 1987 after leaving the job in Dhaka. He tried to do some agricultural work in his village. But due to the necessity of heavy physical work he could not continue.

He then went to Chandina in Comilla District and got a job in a steel factory through one of his relations. While he worked there, his employer was convicted of illegal work and the factory was closed down, so he lost the job and came back home. In January 1988, he got married. His wife is from Burichang Thana of the Comilla District. He took no dowry. His uncles did not like him marrying into a distant village. They advised him to marry into a rich family which would help him get a job or start a business. He was always suspicious of the motives of his uncles - he thought they had been conspiring to grab his property since the death of his father.

He married early because his uncles propagated rumours about his relations with a girl from a neighbouring village. He knew of their conspiracy. He had two younger sisters to get married and he understood that if these stories continued it would be difficult for him to arrange marriage for his sisters. Although he married early, he could not bring his wife to his house for two years due to financial problems. He started to sell some of his dead father's land to keep his family going.

In April 1988, he got a job in a garment factory in Chittagong through his father-in-law. He worked there for four months and was getting a salary of Tk 550 per month. While working in the factory, he secured a job for his wife in the same factory at a salary of Tk 500 per month. Together they earned Tk 1050 per month, which was inadequate to maintain themselves, so they had to get money from their families. In the meantime his wife became pregnant and he sent her home.

After leaving the job in the garment factory, he got a job as an assistant manager in a brick field in Chittagong through one of his relations. He was getting a salary of Tk 1200 per month and free accommodation. His employer did not grant him any leave, so he left the job after 6 months and went home. His wife gave birth to a son. Again he became unemployed and the financial trouble became unbearable with the son added to his family. The owner of the brick field had a clearing and forwarding company in Chittagong Port. The owner was unwilling to give him a job in the company, which provides a larger salary than the brick field, but he begged the employer, explaining his desperate condition and got the job at last. He had work only when ships anchored in the port. When he had work, he got Tk 120 per day which gave him around Tk 2000 per month with free accommodation. He brought his wife and son to Chittagong a few months later. Gradually he realised that his job was risky and his company was involved in smuggling; if he was caught nobody would come to set him free from jail and his family would starve.

Mokhlesur again started looking for a job. One of his relations told him that MCC (Mennonite Central Committee) in Maijdee (Noakhali) was recruiting. One morning, he left Chittagong for Maijdee and found the MCC office. He was worried about how to write the application in English. He thought the foreigners would tear up his application if it was in Bangla. He wrote the application in Bangla and went to a high-school teacher who translated his application. Around 5 months later (in 1990), his father-in-law got a letter (in English) from Faridpur with an appointment. His old father-in-law could not understand the letter very well. Mokhlesur went to the teacher who translated the application to know the details of the letter. He was surprised to find that he would be repaid the cost of going to Faridpur. In Bangladesh it is rare to get any allowance for attending an interview for any job. He managed to get together Tk 1000 (mainly borrowed from his friends) and left for Faridpur.

In Faridpur, he found a hotel and started looking for the MCC office. Next morning, he went to the office for the interview. There were three Americans on the interview board and a teacher from the Bangladesh Agricultural University. The Americans were asking him questions in English and the Bangladeshi teacher translated them into Bangla. During the interview he was offered a cup of tea and he was surprised at the generosity (all these courtesies may have been due to the missionary nature of MCC). After he came out of the interview, a Bangladeshi member of staff took him to another room and told him that he had got the job. He also requested him to keep it secret because local people were trying to get the job and would create trouble if they knew. He was surprised again when an MCC worker gave him the travel costs and a daily allowance of Tk 250 and asked him to start in two weeks. He went home and his family was very happy at the news. He started work at the Faridpur MCC office.

His job at Faridpur was to work with the small partner NGOs. He got training to do his job. Every month, he went home to see his family. His salary was Tk 2575 per month. He worked for MCC at Faridpur for eleven months. After that, he was made redundant due to a budget cut. His immediate superior in MCC, who was a Canadian, was very sympathetic. He advised him to keep in touch and gave him a hearty farewell. Mokhlesur came back to Chittagong and joined his old firm. One day (in 1992), the programme leader of the MCC Soybean Programme sent a man to his village home with a message that there was a vacancy at the Lakshmipur office. In June 1992, he started as the store manager of the Soybean Programme at Lakshmipur. He was getting a salary of around Tk 2900 per month. While working in Lakshmipur in 1992, he saw another advertisement at Maijdee (the Noakhali office). He asked his immediate superiors and they told him that he could apply. The advertisement required three references. He managed to get three foreigners in his office to be referees. He faced the interview and joined the MCC (Mennonite Central Committee) Maijdee office as Homestead Development Facilitator on 1st January, 1993. He now gets a salary of around Tk 5100 per month.

'Why are You Working in this NGO?'

Turning to ask why my respondents were working in their particular NGO, my findings are similar to those of Goetz (1995), that women join the NGOs because they want to use their education and have their own income. Adair (1992) reported that organisations that are most at risk of losing staff in Bangladesh are small NGOs which pay low salaries and are likely to lose their staff to larger local NGOs. Larger local NGOs and international NGOs tend to lose staff to and between each other. This 'transfer of resources' seems to involve men and women equally (Adair, 1992, compare Suzuki, 1998; Fisher, 1997). I have found the same in two case study NGOs, MCC and RDRS (Rangpur Dinajpur Rural Service) but rarely in PROSHIKA and not at all among the SCF 'partner' NGOs. In this section I shall deal with the reasons why the field workers joined their present NGOs.

I have found that the switchover from one NGO to another by field workers depends on the NGO and the field workers' gender. Because of greater mobility, men field workers usually know more about the activities and benefits of other NGOs so they can move to other NGOs easily. Women field workers do not try to think and mix with people of other NGOs like men so they rarely switch over. Apart from local NGOs, most women MCC field workers who had worked in another NGO had been with DANIDA (Danish International Development Agency). DANIDA worked in the Noakhali-Lakshmipur region for more than ten years with a semi-government Bangladeshi organisation called BRDB (Bangladesh Rural Development Board). From the early 1990s DANIDA gradually handed over its activities and resources to BRDB. The DANIDA staff were retrenched, and looked for jobs in other NGOs. Among these, MCC was one of the best choices in terms of salary and benefits and above all, MCC is one of the oldest international NGOs in Bangladesh. One MCC field worker had worked for ASA (Assistance for Social Advancement), which he left due to the heavy workload and lower salary, illustrating the advantages of MCC.

Most field workers who worked in other NGOs before joining PROSHIKA worked for local NGOs and joined PROSHIKA for its reputation. PROSHIKA is one of the largest NGOs in Bangladesh and is known all over the country. On average, 80 per cent of the total costs of the Sakhipur office come from income from microcredit and other sources like renting out irrigation pumps etc. This reveals the level of self-sufficiency of PROSHIKA, which is still improving. Some of its field offices are more impressive than their government counterparts. The PROSHIKA head office in Dhaka is a 14 storey, well-built structure, comparable to that of any big corporation. Because of its reputation and country-wide network, the opportunities for promotion are among the best in the country. One PROSHIKA field worker, now a training co-ordinator, worked for a local NGO for seven years. He regretted this, saying that if he had joined PROSHIKA earlier he could be an Area Coordinator by now. One PROSHIKA field worker had left the Grameen Bank and another had left Ganosaystha Kendra (GK - one of the largest NGOs in Bangladesh) mainly due to conflict with the management and poor job satisfaction (discussed below).

In RDRS, some field workers had been working in other NGOs but joined RDRS (see Table 4.4). They had worked mainly in local NGOs, medium sized NGOs like CCDB (Christian Centre for the Development of Bangladesh), or large national NGOs like BRAC. Those who left the local NGOs did so mainly due to the lower salary, job insecurity and limited area of work of these NGOs. The field worker who left CCDB said it was because of the low salary. The field worker who left BRAC not only did so to get a higher salary but was also fed up with the heavy work load and strict discipline in that organisation. He reported that BRAC is much more committed to microcredit, which puts immense pressure on the field workers. He had to leave home at 6 am in the morning to start work and to work on holidays. If he was even half an hour late for any meeting he had to give an excuse for that to his supervisor. He was living in BRAC accommodation so he was on call to his immediate superiors (field managers) whenever needed, not only in the daytime but even in the late evening. He explained to me the environment in BRAC and told me: *BRAC is not BRAC it is barrack* (Prasun Kumar, Thana Manager, Rajarhat, December, 1997).

These field workers had joined RDRS (Rangpur Dinajpur Rural Service) for its reputation and the better salary and benefits it offers. RDRS has been serving north-west Bangladesh for the last 27 years and has a better reputation than any other NGO in the region. Most field workers told me that the salary and benefits had been much better in RDRS (probably the best in this region) when it was an international NGO (they did not give me the figures, and the NGOs would not).

Table 4.4   Experience of the Field Workers in Other NGOs

Years	Name of NGO								
	MCC Bangladesh		PROSHIKA		RDRS Bangladesh		SCF (UK) 'Partners'		**Total**	
0	31	74	13	81	29	81	15	100	88	81
1-5	8	19	2	13	7	19	0		17	15
6-10	3	7	1	6	0		0		4	4
Total	42		16		36		15		10	
100% *									9	

*Note: Figures are rounded so they do not always add up to 100 per cent.

Source: Field Survey.

The former field workers of SCF (UK) formed their own new NGOs when SCF (UK) pulled out, arguing that they had no alternative. They could only have joined another local NGO, SDS (Shariatpur Development Society), which they regard as a bad NGO. The large national NGOs like BRAC and PROSHIKA had not yet started working in those unions of Naria, so there were no other NGO jobs to go to. Because of their poor education they could not apply for government jobs.

How they Joined their Present NGO: Some Examples

Now I shall describe how some field workers came to work for their present NGOs. All these accounts reveal how bitter the situation was for them to leave their previous NGOs. In the 1970s, NGOs in Bangladesh were mainly involved in social mobilisation and many field workers took their job as a vocation. From the 1980s onwards an NGO job was not a vocation. The activities of the NGOs have changed, and so have the motivation of the field workers. The emphasis on performance has compelled most NGO workers to become more of a professional employee than a social mobiliser.

MCC (Mennonite Central Committee) Bangladesh

1. Rezina Khatun, Homestead Development Facilitator, Maijdee: She passed the SSC in 1978 and worked for DANIDA in its Mass Education Programme. The MCC (Mennonite Central Committee) office was near to her house and her husband regularly went there. When he told her that MCC were going to recruit field workers, she applied, was interviewed and joined in 1988 (March, 1998).
2. Shamsul Alam, Farm Development Facilitator, Companyganz: He passed the SSC in 1972 and worked in farming from 1972-86. He tried to take the HSC examination in 1982 but could not afford to. He saw the MCC advertisement in the newspaper and joined as a night guard. He latter applied for the post of field worker and was successful (March, 1998).
3. Humayun Kabir, Farm Development Facilitator, Companyganz: He took his HSC in 1990, and again in 1991 but failed. In 1992 he did not take it due to financial problems and frustration. In 1993, he passed. When MCC people came to work in his village, he always kept good relations with them. One MCC field worker told him that MCC wanted volunteers for adult education. He applied, was interviewed and joined MCC. While working as a volunteer for six months, he regularly went to MCC and kept good relations with the staff, who told him about an advertisement for an Farm Development Facilitator. He took the written test, was interviewed and got the job (March, 1998).
4. Rokeya Akhter, Women Development Coordinator, Companyganz: She took the SSC in 1993. An MCC field office was near to her house, and an MCC field worker told her brother that MCC wanted field workers. She worked for MCC part-time for six months, and then became a permanent worker as a Women's Development Communicator (March, 1998).

The above examples show that the field workers joined MCC through personal links, and had to start a career because they could not afford further education.

PROSHIKA

1. Abul Bashar, I/C Training, Sakhipur: After passing his BA in 1988 he applied for a job in the Nationalised Commercial Banks, and for the post of Sub-Inspector of Police. To apply for the latter, he paid a bribe of Tk 20,000. Although he did not get the job, he could not get the money back. He joined Ganoshyastha Kendra (GK), a large NGO, in 1988. GK has many operations, including a big pharmaceutical company. During the 1989 flood, the executive director of GK called on all field workers to help save the GK company from the flood. He had to carry sandbags weighing 40-50 kg to save the pharmaceutical company. Many times he fell, got injured and cried. After working in different parts of the country in 1994, he became a Project Manager for GK (Saturia, Manikganz). He married in 1994, but since he was a project manager, he always had to be present at the project site. But the site was in such a remote rural area, that he could not bring his wife there. He visited his wife every week, which his immediate superiors did not like. In 1996, his wife gave birth to a stillborn boy, which so shocked him and his family that he had to stay in Dhaka for seven days. GK issued a letter requiring him to show the cause of his absence from duty and transferred him to Savar. He was devastated by the death of his son and applied to resign from GK. GK took eight months to accept his letter, and only paid him for 4 months. They sacked him and denied him his provident fund of around Tk 100,000. After leaving GK he applied for jobs in the local NGOs like SETU, SSS and CARE. He joined PROSHIKA in 1996 (February, 1998).

2. Abdul Halim, Economic Worker, Sakhipur: After taking his BSc in 1989, he studied law for one year in Dhaka but saw the advertisement for field workers in PROSHIKA and took the job because of what people would say if he was unemployed with a BSc. Now he is satisfied with the job as he is serving the poor, which many senior officials and politicians do not (February, 1998).

3. Lutfur Rahman, Economic Worker, Sakhipur: He passed his BSc in 1989 and became a teacher in a local private school. Although he taught for three years and was paid regularly, the salary was very low and he became a victim of politics among the other teachers. From 1993-94 he went to Singapore. He worked very hard and tried to learn computing, but this proved too expensive. He also applied for several government jobs but without success, which he thinks was due to lack of 'connections' (February, 1998). PROSHIKA was working in his village and its clients told him about it.

4. Shahidul Islam, I/C Area, Sakhipur: His area was flooded in 1989 and he worked for PROSHIKA as a volunteer in its relief efforts. He also tried to join the civil service but could not pass the examinations, and applied unsuccessfully for jobs in the Nationalised Commercial Banks. Before joining PROSHIKA he taught in a private school for six months but left due to the irregular salary (February, 1998).

5. Mohammed Ilias, Economic Worker, Sakhipur: After passing his BSc in 1978 he joined a private high school where he later became headmaster. He was always a believer in socialism and always in conflict with the school management. He was an activist of JSD (a centre left political party). One day in 1981, Quazi Farouq Ahmed, the founder and now the Executive Director of PROSHIKA went to his area (Manikganz) by bicycle and invited him to join, to establish socialism. He immediately joined PROSHIKA without asking about the salary (February, 1998).

6. Sohel Chowdhury, Development Worker, Sakhipur: After passing a BCom in 1993 he joined the Grameen Bank but gradually became dissatisfied with the strict rules in the organisation and conflict with his immediate superiors. One day, his friend came to visit him and his centre manager was very angry, saying, 'Why do your friends come to visit you?' The rule there is that no field worker can leave his station without his manager's permission, even during the holidays. One day, he left his station for home without telling his centre manager. When he returned, his manager was angry again. On another occasion, a big businessman invited all the staff to a feast except his centre manager. When the field staff came back from the feast, their centre manager was very angry with all of them. The field workers replied, if someone does not invite you what can we do? The field worker did not like the system of supervision or the low salary. After leaving, he joined a biscuit factory and worked there for three months. After that he worked as a manager of a hotel in the tourist town Cox's Bazar for around a year but left due to conflict with the owner. He undertook private tuition to support himself and prepared to get admitted to a law college. When he saw the advertisement for an accountant with PROSHIKA he applied immediately. He had to take a written test and be interviewed. When he got the appointment letter as a Development Worker he joined immediately (February, 1998).

These field workers joined PROSHIKA after maltreatment by their previous employers, poor working conditions and failure to get government employment. Some of them joined through personal links. None had the qualifications for the civil service.

RDRS (Rangpur Dinajpur Rural Service) Bangladesh

1. Anwara Begum, field worker, Kurigram Sadar: Her uncle works for RDRS, and told her about the job. She needed money, so she applied, not knowing that she would have to do field work. She thought that the work would be in an office, so she is unhappy with her uncle (December, 1997).
2. Rezwanul Hoque, field worker, Kurigram Sadar: He had applied for jobs in different places - CARE, SCF (USA) - as an employment officer, the navy, Sub-Inspector of Police, Inspector of Narcotics in the Narcotics Department and with the Nationalised Commercial Banks (NCBs) (December, 1997).
3. Mujibur Rahman, field worker, Rajarhat: He went to many factories and met many rich people in search of a job. Before joining RDRS he applied for jobs in more than 150 organisations. His neighbour told him about the advertisement for a job in RDRS when he was working in the Eastern Cable Factory in Chittagong. He left the job in Chittagong because of the low salary and job insecurity (December, 1997).
4. Abdul Quyyum, field worker, Kurigram Sadar: After passing a BA in 1994, he worked for a local NGO; SHODE (Social Human Organisation for Development). He was getting a salary of Tk 2400 per month. He left the job because of the low salary then joined a computer company in Bogra as an operator and was getting a monthly salary of Tk 1800. He left the job because of the low salary. He joined RDRS for its reputation. Before joining RDRS he tried for jobs in a government school, private schools, and as a Senior Instructor in the Sports Ministry. He did not get invited to interview for most of the jobs (December, 1997).

These field workers joined RDRS (Rangpur Dinajpur Rural Service) to escape low salaries and benefits in their previous jobs. Personal links and failure to get government job are also prevalent.

SCF 'partner' NGOs

1. Atia Akhter, field worker, Naria: She saw a circular/notice in the local government office and high-school. Nobody wanted her to do this job. She planned to do the job and continue her education simultaneously,

but, due to her workload, she discontinued her education (November, 1997).

2. Monowara Begum, field worker, Naria: She worked in the relief programme of SCF (UK). Her husband initially objected to her working after her marriage, but her husband was unemployed for a long time, so he accepted her job (November, 1997).

3. Salma Akhter, field worker, Naria: Her nephew saw the advertisement in the Union office. She applied secretly. Her family initially objected to her doing the job. She convinced them that she would work in her village, not go far away and earn good money (November, 1997).

For these women to join SCF (UK) or the later 'partner' NGOs, they had to resist family objections. Personal links are also present here.

Background of the Mid-level Managers

The background of the present mid-level managers are the backgrounds of those field workers who have got promotion. It is usually the good field workers who become mid-level or senior managers (Heyns, 1996; my field work). In this section I shall highlight the criteria which separate them from the other field workers, particularly their socio-economic and educational backgrounds.

If we compare the field workers and mid-level or senior managers of MCC, PROSHIKA and RDRS, the main difference is their educational backgrounds. Most of them hold postgraduate degrees. Above all, they have satisfied their organisations through their services and experience. Backgrounds such as rural middle-class parents and not getting jobs with the government are also prevalent among them (compare White, 1999).

In the case of SCF (UK) 'partner' NGOs the categories which make the difference between the field workers and the directors are skill and family background. The CWDS director is from one of the most influential families in her *char* and is clearly of considerable intelligence.

The Choice of Field Workers by the Senior and Mid-level NGO Managers

This issue was discussed with the mid-level and senior managers of the study NGOs. These are presented in Table 4.5. It seems that three RDRS (Rangpur Dinajpur Rural Service) and one PROSHIKA manager interviewed have become tired of field workers with a BA/MA/MSc. who try to join the government. As MCC (Mennonite Central Committee) field workers are less qualified and less eligible to get government jobs, this problem happens less. The Thana Managers of Rajarhat, Prasun Kumar told me:

> *If you go to the messes of the field workers or agricultural extension officers of less than 30 years of age, or check the drawers of their office desks, you will find guides and notes for civil service examinations* (December, 1997).

The choice made by the SCF 'partner' NGO directors to employ local women seems quite reasonable. They do not like to work with men, and no doubt it is easier for women to work with women clients. Also, women field workers are easier for women directors to control than men.

Why do field managers think graduate or postgraduate degree holders are overqualified? Field workers gave me different reasons. They showed me who got promotion in these NGOs - the graduates and postgraduates. The few HSC holders in MCC, PROSHIKA or RDRS who have reached mid or top levels gave their organisations 20 years or more of work, which compensated for their lack of education. The policy makers or founders of these NGOs cannot ignore the value of their dedication and experience, which are more important than a degree. From the discussion of the backgrounds of the mid-level managers it is clearly the more educated who have reached that level. For RDRS field workers, this creates problems because they cannot work well with the poor (Table 4.5).

Table 4.5   The Preferences for Field Workers of the Mid-level and Senior Managers of NGOs

Who	Qualification	Reasons
MCC Bangladesh		
Both men and women	HSC	Low qualifications, will easily mix with rural people
PROSHIKA		
Both men and women	Not BSc or MSc but rather Diploma agriculturists or engineers	MSc and BSc agriculturists or engineers are over qualified, do not work sincerely and always try to switch over to government or good firms
RDRS Bangladesh (3 managers)		
1. Both men and women	SSC, maximum HSC	Graduates and postgraduates will hesitate to become intimate with the poor and landless and will not bother to walk or ride in remote, inaccessible areas or knee-deep water or during storms.
2. Both men and women but above 30 years	BA/BSc	If over 30, they cannot apply for government jobs so remain serious in RDRS job.
3. Both men and women	HSC	'HSC is enough', no reason given
SCF 'Partner' NGOs		
Women	At least SSC	Can maintain accounts. Men field workers are devious and can cheat

Source: Field Survey.

Conclusion

I have found that field workers are mainly young men and women from middle-class rural families with secondary or higher education. They do not join the NGOs as field workers enthusiastically, but rather to solve their problems of unemployment and poverty. Some NGOs use the family influence of field workers to get access to rural communities. Most field workers have to leave the job when they grow older or (in the case of women) get married. In due course, they will be replaced by new, younger field workers. This cycle will continue into the foreseeable future.

Chapter 5

Field Workers' Personal Lives

The title of this Chapter could be 'Field Workers' Personal Problems' because I have found field workers have many. Most of the discussion below will be on their personal problems, although I wanted to discuss their personal lives. I have found their lives to be full of problems, caused mainly by poor salaries, benefits and working conditions. Most field workers live with them because they have no alternative.

The problems of the field workers differ according to their gender, marital status and age.[1] Obviously the problems of women field workers are more acute, and still more for those who are married (see below). Some of the personal problems are national, and will be difficult for the NGOs to solve. Men and women also spend differently, reflecting their own economic planning, which is mainly influenced by their vulnerability in a very uncertain socio-economic climate.

Fear/Insecurity

Whereas Vladeck (1998) found that the underpayment of nonprofit employees in the US was ostensibly counterbalanced by lower performance expectations and high levels of job security, from my own research the fear of redundancy is the most grievous problem among the field workers of NGOs in Bangladesh (compare Griffith, 1987). Redundancy is a common feature in RDRS and had been extensive in SCF (UK) when it was gradually winding up its operation before the handover, a year before I interviewed them. Although the field workers knew that they would be made redundant in two or three months and get their financial benefits, the immediate shock and financial trouble they faced could be easily ascertained. This is the major difference between NGO jobs and government jobs where tenure in the latter is almost permanent.

---

[1] The ages of the field workers were recorded during the questionnaire survey, I did not record them during my semi-structured interviews.

One SCF 'partner' NGO field worker told me how her colleague started crying when she knew that she had been made redundant. Almost all field workers (except in SCF 'partners') told me that they always live in fear of losing their jobs. In the case of SCF 'partner' NGO field workers, they know why they have formed their own NGOs and that if they cannot maintain high repayment rates their NGO will collapse because their salary comes from the interest on the money they have loaned. Most NGO field workers try to keep good contacts with other NGO managers and field workers (particularly the large and the international ones) in order to get another job if they lose their current one. Redundancy affects not only field workers and ancillary staff, but mid-level and senior managers too. When I returned to Dhaka to leave for the UK after my field work I heard from the SCF (UK) Dhaka office that the accountant and the administrative officer in the SCF (UK) Shariatpur office had been made redundant in March 1998.

Due to this job insecurity, all field workers try to save. The women are more successful as men have to spend more on their families. Economic dependency is very high in Bangladesh. It is usually the men who have to support not only their nuclear family but in many cases their parents, younger brothers and sisters etc. Most women field workers usually support their family's income rather than provide it.

Financial Difficulties

This subsection may seem surprising when the field workers are themselves engaged in the alleviation of their clients' poverty, which is the more acute. The poverty of the field workers still deserves mention, particularly when the field workers of SCF 'partners' have faced a drastic fall in their salary: around 70 per cent since the handover. All SCF 'partner' NGO field workers told me how difficult it was to cope. Most married field workers were told by their husbands to leave a job with such a poor salary. One field worker told me that she is doing the job for low pay because she is a woman. If she were a man, the SCF (UK) managers would expect trouble from forcing men field workers to accept a salary reduction. She thinks women are more obedient than men. A lower salary means lower purchasing power. All these field workers told me that since they still work and were earlier getting a good salary, their parents and relations still expect help and gifts during the festivals (the Eids). With their lower salary they are finding it very difficult. To satisfy relatives with gifts, most field workers start saving a couple of months before a festival, which makes their lives even harder. The poverty of the field workers of the SCF 'partner' NGOs may be illustrated by the situation of Khaleda Begum.

Khaleda's husband is an accountant at the Income Tax office in Dhaka. With his low salary he cannot afford to keep her in Dhaka. He lives in a mess and comes to visit her every week which is also very expensive. She joined SCF (UK) in 1992. Khaleda was getting a salary of Tk. 4905 per month before the handover. She thinks it was a good salary for her. She could save some money and helped her husband to pay the fees to take his MA examination. Unfortunately, he did not pass the examination. When she married, her husband was unemployed. To get a job in the Income Tax office her husband had to pay a bribe of around Tk 40,000; she paid Tk 20,000 and her family paid the rest. While working in the SCF (UK) she invested around Tk 30,000 in the stock market from her saved money. Unfortunately she lost all the money due to a crash in the market. Now she gets Tk 1500 per month. Her husband is trying to go abroad but she cannot help him any more because she cannot save from her salary of Tk 1500 (November, 1997).

The problem of financial stress also exists in the lives of the field workers of other study NGOs.

Gita Rani (married), Homestead Development Facilitator, MCC, Maizdee: Her husband is a teacher at a private primary school but does not get his salary regularly. She has two sons and a daughter. Her husband and her children live in Maizdee town because the schools are better there. She cannot live in Maizdee due to the rules of MCC, so she lives in a rented room in her working area. Her children do not like to be with her even at weekends because it is a shabby room with no electricity, and a shared kitchen and toilet. So, every Thursday evening she leaves for Maizdee and returns to her working area on Sunday morning in tears. She gets a salary of about Tk 5000 per month and to live in the rural areas she spends around Tk 2000 on herself. The rest of her salary is spent on her family. She told me that she is doing the work because if she had no income she might have to stop her children's education. Although both she and her husband earn money, every month they have to borrow particularly when her husband does not get his salary (March, 1998).

Qamrul Islam (married), I/C Training, PROSHIKA, Sakhipur: He joined PROSHIKA in 1990. He has three younger brothers and two younger sisters. His father died in 1992 and he has to defray the costs of his brothers' education. The next brother is an MA student at Chittagong University, the third one studies at Jagannath University College in Dhaka and the youngest passed the HSC in 1998. The elder two brothers do private tuition to support themselves but it is inadequate. Qamrul gets a salary of around Tk 6000 per month but has to pay at least Tk 2000 per month for his brothers' education. His two younger sisters have reached marriageable age. He does not know how he will arrange their marriage. He told me that the only way out is for him to sell land and use his savings in the provident fund. He saves Tk 200 per month in a deposit pension scheme. He married in 1995 at the age of 32, although he was reluctant to, for financial reasons. But his mother and uncles married him off because they thought he was getting past the right age. His wife has a BA degree and she looked for work in government primary schools, the Food Directorate, the Postal Department, and the Youth Development Directorate but she did not even get an interview for these jobs. He thinks his wife did not get an interview because he has no 'connections'. Recently his wife has got a job as a Development Worker in PROSHIKA, which he thinks is due to him. He did not want to allow his wife to work with PROSHIKA due to the high workload and low salary. He has reluctantly agreed, due to their financial problems. He has a daughter aged two and a half years. He is worried about who will take care of her. Above all, unless they (he and his wife) get a posting in the same Area they will face the problem of family dislocation and higher maintenance costs. This will make his life even more difficult (February, 1998).

I agree with Rahman and Islam (1994) that field workers should be motivated by financial rewards. The field workers' relative poverty, however, is not only caused by low salaries, but it is exacerbated by the rule that they have to stay in their working areas. Due to family dislocation they have to visit their families regularly, which also involves costs. (The problem of family dislocation will be elaborated below). Teachers, for instance, often earn less but have a job which is less stressful, of high social status and not subject to family dislocation.

Where is Home?

A major problem for almost all NGO field workers is their accommodation. Only SCF 'partner' NGO field workers did not complain to me about their accommodation. This is quite natural, as they live in their own areas and work in their own or neighbouring villages. The three other study NGOs do not allow (as a rule) their field workers to work or sleep in their own village. This causes the problems of family dislocation (discussed below) and of accommodation. In that sense, the field workers of SCF 'partner' NGOs are rather lucky.

As a rule NGO field workers have to live in their working area. NGOs do not allow their field workers to live in the office accommodation, except for a few days or with special permission from their superiors in exceptional cases. In PROSHIKA it is not allowed for more than one month except for women workers who cannot find accommodation. So, in general, field workers cannot stay in their office accommodation. However, no field workers as they told me, want to live in the office accommodation because they could be called on at any time for work (they all cited the example of BRAC). I found a strong sense of discontent among the MCC field workers over the rule which compels them to live in their work area, as it causes family dislocation and problems like education of their children, insecurity etc.

A major problem for field workers is to find suitable accommodation. Most field workers told me that people in rural areas do not welcome outsiders and look on them with suspicion. This is very familiar from my experience in rural Bangladesh. If this problem is overcome, then comes the problem of finding suitable accommodation with a good latrine and cooking facilities and adequate security. If a house is available, next comes the question of affordability. On average, field workers interviewed have to spend 10 to 25 per cent of their income on accommodation. This might seem acceptable, but many field workers pointed out to me that they have to spend money for two homes - one for living in their work area and another for their families in a nearby town or home village. About 10 per cent of MCC, 30 per cent of PROSHIKA and 20 per cent of RDRS married field workers in my studies live with their families in their work area. Although the rate is highest among the PROSHIKA field workers, most of them are couples who both work for PROSHIKA. Otherwise those who live with their families are mainly men and their wives are housewives. Most unmarried men field workers live in a mess or rented room while women live in a rented room, or if possible in a relation's house. There is a severe scarcity of housing in the rural areas, but the urban areas are no different. Usually a newly transferred or appointed field worker gets help from his or her colleagues to find accommodation. Sometimes they approach local leaders or clients for help. The reputation of the NGO and the departing field worker in many cases work as decisive factors in getting accommodation. Sometimes new field workers find it very difficult to find accommodation if relations between the previous field worker and local leaders and clients were poor.

The problems of unmarried field workers are different. For unmarried men, the major problem is that many rural families do not like an unmarried man from outside to live in their village for reasons of privacy.

No-one who decides to rent out a room to an unmarried man will have grown up daughters. Sometimes they have to face criticism from their neighbours who have grown up daughters or who do not like an outsider in their village. For unmarried women the problem is more acute. They cannot live anywhere without taking care of their security. Even if a woman gets a house to rent she has to check how secure she is. Many women told me that their assumptions proved wrong when they were disturbed even by older people, which they did not anticipate. So, the choices for accommodation for women field workers are even more limited. Some women told me that one way out is to live in the houses of influential families. But here again if someone from that family starts creating trouble then the situation becomes worse.

An opportunity that many women field workers try to take is to live with their relations. If they are lucky, they usually do not have to pay any house-rent, but as a custom they have to give gifts to the members of the host family. Sometimes they have to buy costly items like fish and chicken. Most field workers told me that these cost more than the house-rent they might pay. Still, women field workers prefer to stay with their relations than to live with unrelated people because they feel more secure.

Below, I illustrate the accommodation problem in some detail.

MCC Bangladesh

1. Mohammed Humayun Kabir (married), Farm Development Facilitator, Companyganz: He joined MCC in 1995 and was always posted in the *char* area. Due to the rules of MCC he had to live within 3 km. of his work. Being an outsider, he had to get permission from the local leaders to take a room in the local bazaar (market). There he faced many problems like unhygienic food and open latrines, so he had regular stomach upsets. After six months he convinced a local Union Parishad member to provide him with a room in the local primary school where there was a better latrine and a tube-well. Now he lives for free in a cyclone shelter but eats in a hotel which is not hygienic and is expensive. He cannot manage time to cook or afford to bring his family (March, 1998).
2. Rokeya Akhter (married), Women Development Coordinator, Companyganz: She has to live in the *chars*, because it is her working area. She lives with her one and a half year old daughter. It is very difficult to find a good accommodation in the *chars* with her income. Her frail house is made of thatch and starts swinging in any strong wind (March, 1998).

PROSHIKA

1. Sohel Chowdhury (unmarried), Development Worker, Sakhipur: When he was working for the Grameen Bank at Manikganz he slept on the tables in his office for around a month. After that he moved to a mess with one of his colleagues. He had to pay Tk 200 per month as house-rent while his salary was Tk 1600. Now he gets a salary of around Tk 2870 per month and lives in a rented room for which he has to pay a house-rent of Tk 300 (February, 1998).

RDRS Bangladesh

1. Prasun Kumar (married), Thana Manager, Rajarhat: If he lives in government staff quarters, as per rules he would have to pay 40 per cent of his basic salary as house-rent, which he cannot afford. He lives in staff quarters of the local post office as a sub-let and pays Tk 1000 as rent. This is one km away from his office. Mr. Prasun thinks that the residence of the field staff should not be in the office compound because their immediate superiors can call on them at any time, even at midnight, as he experienced in BRAC (December, 1997).
2. Anjoli Roy (unmarried), field worker, Kurigram Sadar: She lived in the house of the uncle of another RDRS field worker when she was posted at Lalmonirhat where she had to pay Tk 600 per month for food and accommodation. Now she lives in a room where she has to pay a house-rent of Tk 500. Her house is one km from her working area and she has a bicycle. Her total salary is Tk 3040 per month of which she sends Tk 1000-1500 to her family. She told me that she does not feel as insecure in her present accommodation because she lives in the house of an influential man and her clients are not far away, but still she never stays out after dark (December, 1997).
3. Rezwan (married), field worker, Kurigram Sadar: He lives in a mess with one RDRS field worker, one BSCIC (Bangladesh Small and Cottage Industries Corporation) employee and one BRAC field worker. He gets a monthly salary of around Tk 3300 and has to pay Tk 350 as house-rent and Tk 30 for electricity. He cannot afford to rent a larger house to bring his family (he has a son) to live with him (December, 1997).

Problems of Family Dislocation

Goetz (1997) found that working in rural areas poses personal problems for both women and men field staff, because it means that staff must move far from their homes and adjust to a new environment. If they have families, there are problems associated with moving a family, finding new schools and finding accommodation. Frequent transfers can exacerbate these problems. The ways in which the expenditure of the married field workers go up due to family dislocation have already been discussed. Montgomery *et al.* (1996) found family dislocation to be a major reason for high staff turnover in BRAC, but from my research many field workers cannot leave their jobs because they have no other jobs to go to. So, they have to accept this family dislocation. As one SCF (UK) mid-level manager (Saiful Islam, Accountant) told me about his feelings after working for 7 years at Naria leaving his family in Kishoreganz:

Now if I get a sentence of 7 year harsh imprisonment I can bear it (November, 1997).

Some field workers never take their families with them because of the difficulty in finding accommodation, schools for children etc. For example; Abdus Salam, Assistant Thana Manager, Rajarhat joined RDRS in 1980. Over the last 18 years he never lived with his family. In some years he was transferred more than once. He kept his family at his village home and stayed in a mess. He thinks this was the best approach (December, 1997).

Apart from the family security and children's education, field workers face several problems due to family dislocation. Although some senior managers told me that they consider field workers' applications for convenient posting, I heard several complaints about posting.

The way family dislocation affects the lives of field workers could be understood from the following examples.

Lufunnahar Begum, Homestead Development Facilitator, Maizdee, MCC: Her daughter studies in class Ten at Maizdee. She visits her daughter every weekend. She is always worried about her daughter's education and health. Her daughter very often tells her to leave the job and to live with her. She told me about five MCC field workers who left their jobs due to family dislocation and some who are planning to do so (March, 1998).

Abdul Awwal, Thana Manager, RDRS, Kurigram Sadar: His wife is an Assistant Statistician at Thakurgaon Thana Health Complex and they have two children. Every Thursday afternoon he leaves Kurigram and reaches Thakurgaon at around midnight. Next morning (Friday) he undetakes the weekly shopping for his family. In the afternoon he leaves Thakurgaon and reaches Kurigram at midnight. The transport system in Bangladesh is erratic and he and his family suffer continual stress. A few months ago there was a burglary at his house in Thakurgaon. The dacoits took away all the ornaments, valuable clothes and a tape recorder from his house at gunpoint. For two years he had been asking his superiors to transfer him to Thakurgaon or nearby. After the burglary he became fed up with this life and is planning to resign to try to start a business. He gave me examples of other field workers getting a posting at their desired stations very easily because of their 'connections' (December, 1997).

What is Different for Women Field Workers?

The problems of the women NGO field workers are more severe than those of their male counterparts. Woolcock (1998) reports a 26 per cent turnover within three months of training in Grameen Bank, and even higher rates among women. Those who survive this period will likely have a career spanning several decades of living under difficult circumstances in isolated settings, often away from family members and loved ones. I found the same in my study NGOs, except the SCF 'partners'. Goetz (1997) found that both men and women NGO field workers were exhausted by field work, which resulted in health problems and claims of fatigue and physical strain. My work confirms her findings. I have found the same and would add that these problems affect women more seriously due to their poor nutritional status, identified by Goetz (1997).

Goetz (1997) also recorded women's hygiene and sanitation problems, leading to gastroenteritis and other intestinal and liver infections. The paucity of latrines and difficulties in managing menstruation causes women to restrict their fluid intake which results in dehydration and urinary tract infections (Goetz, 1997). As for Goetz, all women field workers told me that it is no use reporting these problems to their immediate superiors (even women) or their family. The reply would be either to leave the job or to live with it.

Ackerly (1995) reported on the special arrangements provided by SCF (USA) for their women field workers for the collection of money, special secure accommodation and living and sanitation (Ackerly, 1995). From my surveys, interviews and observations, no women field workers where I worked enjoyed such privileges. Goetz (1995) found that unmarried women field workers face criticism from neighbours, colleagues and relatives that they work outside, losing personal honour and integrity. Married women field workers faced criticisms not only from relatives and colleagues but from husbands for mixing with non-related people, giving less priority to domestic work and, above all, being engaged in work which is not 'respectable'. My findings confirm this. I would like to add a comment by a RDRS field worker (Hasina Begum, Union Organiser, Ulipur) which indicates the severity of women's problems:

> *We have to work in rain, in cold, at all times. If a woman's saree gets soaked by rain she cannot walk because of shame. If a man slips on the muddy road people would laugh, sometimes show sympathy. If a woman slips people would not only laugh they would ridicule her and sometimes say, 'Why has this woman come out of her home? How shameless she is!'* (December, 1997).

All women field workers told me of their problems of personal safety. Particularly, they faced hostility from local leaders, both powerful and religious when they started their work (except for some SCF 'partner' NGO field workers, because of their family influence). Eickelmen and Piscatori (1996) found the same in BRAC.

The security of the women field workers is always a problem and all field level managers I interviewed complained to me about the difficulties this makes for them especially when field workers must carry money because they are involved in microcredit. All women field workers told me that carrying money is not safe for men either. Another problem is that people (and also criminals) know the dates for collection and distribution of money. This is a real problem because law and order is poor in rural Bangladesh.

Apart from the problems of safety in work women field workers always face insecurity at home which is exacerbated when they do not live with a man, particularly their husband (discussed below). All women field workers except SCF 'partners' told me they feel insecure as neighbours think that they are well-off because they get a handsome monthly salary. Most of them told me that they do not draw their full salary all at once, and none keep money in the same place due to the fear of burglary.

All women field workers (except those of SCF 'partners') told me of initial problems of teasing by youths, who would damage their bicycle or create obstacles on their way. Gradually these problems are reduced, or in a very few cases have disappeared.

Married Women Field Workers

For women, problems of family dislocation are more severe, but affect single and married women differently. Goetz (1997) discussed both family dislocation and women's inability to give time to domestic responsibilities so that married women field workers have to pay for maids and childcare, which increases anxiety (Goetz, 1997). I found the same problem among the married women field workers. Transfers create serious problems for all of them, except in the SCF 'partner' NGOs.

A major problem for married women field workers is their husbands' objection to their jobs on pretexts such as neglect of housework or child-care, interaction with other men etc. Other marital problems mentioned to me by married women include: due to their workload they cannot take care of their in-laws; their husbands may be ridiculed by their friends and relatives for being so poor that they allow their wives to work in the field, to meet many men and to have male superiors; they must ride a bicycle or motorcycle. Some 62 per cent of the married women field workers told me that their husbands explicitly dislike them doing their job. Some told me that they try to buy the favour of their husband and husband's family through spending money on their in-laws, helping their in-laws financially or giving gifts. For example:

Rokeya Akhter, Women Development Coordinator, MCC, Companyganz: Her husband is a field supervisor for the Social Welfare Directorate. When she was married she told her father-in-law that her salary was Tk 2200 per month, when she was actually getting Tk 3000 After her marriage she began to contribute Tk 500 per month to her husband's family. Soon after the marriage her father-in-law told her that their family will depend on them for an income. She continued to contribute Tk 500 for one year. The lady who did the match-making for her marriage was her husband's colleague. One day the match-maker went to the MCC office and came to know about her exact salary. Her husband and his father became very angry at her deceitfulness and threatened her, saying she must either leave the job or pay more money to their family. She apologised and now gives Tk 1500 to her husband's family per month from her salary of Tk 4480. She has to live in the char in her working area where she has to spend at least Tk 1500 (March, 1998).

There are worse cases. For example:

Lutfunnahar Begum, Human Development Facilitator, MCC, Maizdee:
She married a homeopathic doctor in 1984. In 1985 she joined MCC as
an extensionist. Her husband and his family became very angry at her
taking the job and in 1986 her husband divorced her. Now she lives in
the village where she works and her daughter is growing up with her
parents at Maizdee. Her husband does not even contact his daughter
(March, 1998).

Another major problem for all married women is that of child-care and
children's education. Although the latter affects both men and women,
women field workers seemed to me more worried about it saying that their
husbands very often blame them for not being serious about their children's
education due to their workloads. All women NGO field workers who have
children told me of their problems with child-care. Among the women field
workers with children in the survey 36 per cent depend on a maid, 29 per
cent on their mother, 29 per cent on other relatives and six per cent on their
husband. When their children are taken care of by a maid, women are
always worried about safety and proper care. Finding a suitable and reliable
house-maid is also another problem. When relatives take care of their
children women very often continue to worry because relatives may not be
very responsible and may have other commitments. All married field
workers, men and women, expressed to me their grievances about the
bringing up and education of their children. Because of their heavy
workloads and their domestic responsibilities, the women cannot attend to
the education of their children in the evening, which is highly desirable
given the poor standard of education in rural schools.

　　As mentioned earlier, it is difficult to find a good school in the remote
rural areas where field workers have to live. Not only the field workers; one
woman Programme Leader of MCC's Homestead Resource Development
Programme expressed concern about her son's education, even though she
lives in a district town (Sinjita Alam, March, 1998). One MCC Women
Development Coordinator expressed the problem to me:

I have ruined my life by working as a field worker. Now I am ruining
my childrens' lives (Zakia Begum, Maizdee, March, 1998).

Among married women field workers in the survey, 58 per cent live apart
from their husbands. In most cases, their husbands come to visit them
weekly or monthly or sometimes the women themselves go to visit their
husbands and children. Living without husbands of itself creates problems,
such as lack of security, financial problems etc. Some married women field

workers talked of problems like shopping for daily necessities for which they have to depend on maids or other people. Due to the workload, very often they cannot do the shopping because fish and vegetables are available only during fixed hours in rural markets.

The extent of difficulty in the lives of the married women field workers is elaborated in the examples below:

1. Mita Rani, Homestead Development Facilitator, MCC, Maizdee: She always finds it difficult to reach her work area in time in the morning due to her domestic responsibilities. As per the requirements of MCC she has to complete many forms. She cannot do them in the field or office due to her workload but must do them at home in the evening, which affects her child-care, cooking and other domestic responsibilities. Very often her husband becomes irritated and sometimes tells her to leave the job. She told me with grief that she has to listen to many harsh words from her superiors in the office for being late or inefficient and from her husband at home for lack of commitment to domestic responsibilities. Sometimes she loses patience and thinks that she will leave the job (March, 1998).

2. Razia Begum, field worker, Rajarhat: She joined RDRS in April 1987. When she joined RDRS her relatives and neighbours were very critical, saying that she should not be working for the Christians. Her father became very frightened initially. After joining RDRS she did not ride a bicycle but walked with it for around six months because of shame and ridicule from other people. She thinks the situation has improved for women field workers in the last 3-4 years because women field workers on a bicycle face less ridicule from other people. During her marriage her husband's family has often asked her to leave the job. She and her father did not yield. Soon after the marriage her husband asked her to leave the job. He no long says that, but still on occasion does not hesitate to express his dismay. Her mother-in-law respects her because of her job. Every morning Razia cooks for her family (she has two daughters). She finds it very difficult to cook and finish the household chores and leave for her work in time. She told me that she feels guilty that she cannot take proper care of her children. She takes biscuits or puffed rice as a mid-day meal. Her sister-in-law cooks the lunch for her family. She told me that her sister-in-law does not like to cook for her family and very often creates trouble out of envy saying that she cannot manage the time and her own family suffer. Razia Begum cannot return home before 5pm. She told me that she was always worried about her job and her children (December, 1997).

Although NGOs say that they give special considerations to the problems of women field workers, I have found many exceptions too. One MCC Women Development Coordinator told me that when she was pregnant she had to travel 18 km (eight km by Tempo, seven km by rickshaw and three km on foot) from home to work every day. She told me sorrowfully that her immediate superior, who was a woman, was not sympathetic to her problems. The superior sometimes did not hesitate to mistreat her on some occasions too. This example surprises me as MCC tends to set high standards for the way staff are treated.

Even so, all women field workers interviewed told me that they would continue their jobs as long as they could. They think the job has given them independence and economic and social status. Most women field workers try to keep the exact amount of their salary secret and tell a lower amount to their husbands and family, which helps them to save money. Some husbands go to their wives' offices and try to find out their exact salary. If they succeed it creates family trouble (as mentioned above).

On the other hand one field worker helped her husband to start a business and another invested her savings to expand her husband's business. As mentioned above, one SCF 'partner' NGO field worker helped her husband to take his MSc examination. These are a few examples of how the employment of women field workers helps to improve their status and the condition of their families, for which NGOs should be lauded.

Problems of Unmarried Women Field Workers

Goetz (1997) found that for single women field workers living far from home means the loss of security and support provided by the family. I would make the same observation. She also found that it becomes difficult to begin arrangements for marriage, as families are not close enough to consult with daughters and introduce them to prospective partners. Some women cited to Goetz the need to get married as their reason for leaving BRAC (Goetz, 1997). I would make the same comment after talking with the unmarried field workers of my study NGOs but I have found some exceptions too. I have found that employment in NGOs helped some unmarried women field workers to get married because young men want to marry working women for economic solvency. Marriage between working field workers is not uncommon in PROSHIKA and after marriage they can ask for postings in the same area. Some mid-level managers of PROSHIKA, RDRS and SCF (UK) proudly told me how they did the match-making or helped in arranging marriages for unmarried women field workers. Some feel it is their social and religious obligation while some were requested by parents of the field workers to do it.

Still, it is common that unmarried women field workers face problems in getting suitable husbands (compared to women, unmarried men field workers face no real problem in getting married). In Bangladesh, a man qualifies as a good groom when he has a respectable job or a regular income. If women marry, they have to convince their husbands to let them continue with their job, although this becomes very difficult when he works far away from her work area. I also found that some women field workers save money to pay their dowry (i.e. money for going abroad, bribes for a job, furniture, electronics etc.) or to bear part of the cost of their marriage. This is interesting, since most women field workers work for the elimination of the dowry among their clients. Most unmarried women field workers expressed to me their anxiety about marriage, saying that it is very difficult to find a suitable groom. Whenever a suitable groom is available then comes the problem of a dowry. This problem is more severe in the case of Hindu women field workers.

Risks for Field Workers

There are many places in rural Bangladesh to which field workers cannot bring their families due to poor living conditions, paucity of schools etc. Those married field workers of MCC, PROSHIKA and RDRS who do not live with their families told me that they are always worried about their families and children. MCC field workers are lucky compared to those of RDRS and PROSHIKA where very often field workers have to work at weekends and cannot visit their families. SCF 'partner' NGO workers have no family dislocation problems because they can live in their own village.

Not only women, but most men field workers reported to me the problem of accidents during work. This is more common among the MCC, PROSHIKA and RDRS men field workers who ride motorcycles on the dilapidated rural roads. Although PROSHIKA women field workers drive only 50cc motorcycles they are not free from risk. Those who ride motorcycles told me that due to the workload they have to drive fast and in many cases they face accidents to save careless children, aged people, livestock or poultry. However I have found that some field workers unnecessarily speed and do not wear a helmet. I have found the MCC rules more sensible and strict on speed limits and the use of helmets. Among those who ride motorcycles there are very few field workers who have not had an accident.

The physical problems faced by the field workers have been discussed above. Apart from the problems mentioned earlier, most field workers told me that they face irregularities in their meals. When they have to eat outside the home, they have to eat in restaurants which are expensive and unhygienic. Even where restaurants are available, women field workers in many cases cannot eat due to lack of privacy, so they must go hungry. Some women field workers take food with them. Those married and unmarried men field workers who live in a mess either cook for themselves (for which it is difficult to manage the time) or depend on a part-time cook. Other problems like accidents, lack of security (particularly carrying money), or working in certain areas where dacoits and outlawed militant political parties are active pose great risk to the field workers.

Expenditure Patterns of Field Workers

Goetz (1995) found that more men than women said their income was devoted to the full support of their spouse, children, and in-laws. However, more women than men sent their incomes back to their natal families to support their parents or siblings. Most married women use their incomes to supplement their husband's income. I found the same in my study areas.

In the last Chapter I discussed the future plans of the field workers. I found that women field workers have more propensity to save or invest money in savings schemes or buy ornaments, which bring both economic and social prestige. Compared to men, most unmarried women field workers buy ornaments or save money which will help their parents in fulfilling their marital obligations. Married women field workers can save more than men because women usually supplement their husbands' income. Usually men are considered as the breadwinners and they have to shoulder the responsibility of running the family.

Table 5.1 Expenditure Patterns of Field Workers

Unmarried Man
  Supports himself
  Sends money to the family
  Bears the cost of education of younger brothers and/or sisters
  Buys gifts during home-visit or festivals
  Saves money in the credit union of their NGO (MCC only)
Married Man
  Supports family
  Supports parents
  Bears the cost of education of younger brothers and/or sisters
  Saves money in the credit union of their NGO (MCC only)
  Mortgage in agricultural land
Unmarried Woman
  Saves money in deposit schemes
  Buys her own clothes, cosmetics
  Gives gifts to parents, relations
  Helps brother to go abroad
  Buys ornaments
  Buys electronics (transistor, tape recorder etc.)
  Repays security money (RDRS only[2])
  Saves money in the credit union of their NGO (MCC only)
  Bears the cost of education of younger brothers and/or sisters
Married Woman
  Saves money in deposit schemes
  Invests money in husbands' business or bribes for getting a job or pays
    examination fees
  Children's milk, food, education, clothes
  Buys ornaments
  Buys cattle
  Buys share/savings certificates
  Buys cultivable land
  Repairs or constructs house
  Bears the cost of education of younger brothers and/or sisters or going abroad
  Saves money in bank (in fixed deposit)
  Saves money in the credit union of their NGO (MCC only)

Note: Of the 109 respondents, 18 per cent were unmarried men, 37 per cent were married men, seven per cent were unmarried women, 37 per cent were married women and one was a divorced woman.

Source: Field Survey.

---

[2] RDRS field workers have to deposit Tk 10,000 to join.

Compared to urban middle-class women, as found by Islam (1995) field workers, who are mainly from rural middle-class families, could not afford to keep a major part of their income for their own use. Mainly due to the financial difficulties of their families, field workers have to help their husbands, in-laws and parents and bear some costs for their children. This difference shows the disparity in the situation of women in urban and rural areas, where the latter are the majority. I agree with Islam (1995) that the employment of women is breaking the norm that the 'son is an investment for future income'. This is a big achievement and NGOs should be lauded for that.

Table 5.1 gives a vivid picture of the expenditure pattern of the field workers of NGOs and clearly shows the differences in the expenditure pattern of men and women. The men (particularly married) field workers' main responsibility is to support their families, where most of their income is spent (very often it is inadequate). Since women field workers usually supplement their families' income, it is easier for them to save or invest or help their family. The employment of women field workers has therefore helped them to increase their economic self-sufficiency and their status in their families. At the same time this has dissatisfied many men field workers who say that if a man were employed instead of a woman, this could help a family. All NGO field workers save money or are forced to save money in a provident fund due to the rules of their organisations. In times of need field workers can release money from their provident fund. One PROSHIKA man field worker told me that he bought some essential furniture from the money in his provident fund. The presence of a credit union in MCC is really laudable and all MCC staff appreciated the system. The expenditure pattern of the field workers clearly demonstrates their way of living in a very uncertain professional and personal environment. It also shows how the employment of women affects their economic and social status. From this research, the positive effects outweigh the negative.

Case studies of individual expenditure patterns follow.

MCC Bangladesh

1. Mita Rani (married), Homestead Development Faciliator, Maizdee: She gets a monthly salary of Tk 5552. Her husband is a block supervisor in the Agricultural Extension Department of the government, with a salary of around Tk 4000. She saves Tk 200 in a deposit pension scheme for herself, Tk 300 for her son and Tk 200 for her husband. From her savings she has bought a little agricultural land with Tk 22000. Interestingly, she has intentionally bought the land in the name of her

husband. When I asked her the reason for this she said 'Why not?' (March, 1998).

2. Kalpana Rani (married), Human Development Facilitator, Maizdee: Her husband is involved in farming. She gets a salary of around Tk 5854 per month and has saved around Tk 10000 in the post office. She helped her husband in partially paying his costs for going to Abu Dhabi (Tk 40.000). She also helped her husband during the marriage of her brother and sister-in-laws (around Tk 20000). She saves Tk 200 per month in the credit union of the MCC (March, 1998).

3. Zahid (unmarried), Partnership in Agricultural Research and Extension Supervisor, Maizdee: He gets a monthly salary of around Tk 5000. He has three younger sisters who are unmarried and are studying. He can barely send Tk 2000 or 2500 per month to his family after supporting himself. He does not save in any pension scheme and has no plan to marry before the marriage of at least his two younger sisters (March, 1998).

4. Shamsul Alam (married), Farm Development Facilitator, Companyganz: He has six children, the first son is a BA student, his second son is studying in class eight and the youngest son is four years old. His eldest daughter passed the SSC in 1997 and has discontinued her education. The second daughter is in class ten and the youngest in class four. He gets a salary of Tk 5807 per month and has to spend at least Tk 2000 on living in the field. He sends the rest to his family. His only saving is of Tk 100 in the credit union of his NGO (March, 1998).

RDRS Bangladesh

1. Abdus Salam (married), Assistant Thana Manager, Rajarhat: He joined RDRS in 1980. He has three daughters and a son. Last year he took out a loan of Tk 32000 from his provident fund for the marriage of his eldest daughter. In addition to that he had to mortgage out some agricultural land for Tk 20000. He still has to buy a show-case, a bed and a dressing table for her daughter. He told me that he did not know how he would manage the money for this furniture. He has no more agricultural land to mortgage out. He has been serving RDRS for 18 years but could not buy a piece of land. The land and the house he has are inherited from his father (December, 1997).

SCF 'partner' NGOs

1. Masuma (married), field worker, Naria: She gets a monthly salary of Tk 1500 of which she saves Tk 500 in a deposit pension scheme and spends Tk 500 on her two year old daughter. She invests the other Tk 500 in her husband's business; a saree store in the local market. She believes her husband's business has expanded through her investment. She has bought two cows (Tk 2000 each) from her saved money which will start giving her a regular income soon. She has also bought ornaments costing around Tk 10000. She told me that she could do that because she saved more before the handover. She knows she will have to leave her job when she grows older (November, 1997).

2. Salma (unmarried), field worker, Naria: Before the handover, she was earning around Tk 4500 per month and could save around Tk 2000. She has bought ornaments worth around Tk 10000. She also gave her brother Tk 30000 to go to Dubai but he was cheated by his manpower agent and returned penniless. She is afraid that her brother will not be able to repay her money, but she is proud that she could help her brother and hopes he will remember that. Now, in the new NGO, she gets Tk 1500 per month, and saves Tk 300 in a deposit scheme and she has to spend Tk 200 on transport. She saves Tk 400 in a bank and spends the rest on shopping once or twice a month for her family (November, 1997).

Field Workers' Recreation

In order to get a clear picture of the lives of the field workers I tried to find out how the field workers spend their free time. I found that PROSHIKA field workers have the longest working hours, followed by RDRS. I have mentioned earlier that only MCC and SCF 'partner' NGO field workers can enjoy weekends. So, the opportunity to spend time with family and undertake recreation is highest among the MCC and SCF 'partner' NGO field workers.

There is an interesting gender variation in the recreation patterns among the field workers. The SCF 'partner' NGO women field workers enjoy TV, although there is no electricity in the *chars*. Televisions are run on batteries and their prime time is from 8 pm to 10 pm when all the soaps and serials are shown. Those women who have TVs, or can get a seat in their neighbourhood, can enjoy this opportunity. Still, most women NGO field workers have very little time to watch television. This is more difficult for married women whose top priority is to fulfil their domestic responsibilities. Men field workers of MCC and RDRS told me that their recreation is not only TV (if possible) but also cinema (particularly in RDRS). Due to safety problems and domestic responsibilities, women field workers rarely go to the cinema. Also, men field workers can spend their time playing cards. Playing cards is easier for men field workers who do not live with their families.

Except SCF 'partner' NGOs, the study NGOs organise an annual picnic. Every year RDRS organises a badminton tournament for its staff. Most PROSHIKA field offices have TV sets for the trainees. Field workers can watch television with the trainees in the evening if they can find the time. Everywhere there is a problem of power cuts which may deprive them of the satisfaction of watching. But still, many men (let alone women) field workers cannot watch TV or go to the cinema due to the workload, paucity of electricity and lack of a suitable environment. I would like to sum up the recreation life of the field workers from the following statement by Bimal Kumar, an RDRS Agricultural Extension Worker at Ulipur:

I leave my mess at 7.30 or 7.45 in the morning. I come back at 7pm or 8pm. Then I find that there is a power cut and usually it is not restored before mid-night. So, I take my dinner with the help of hurricane or oil lamp. After dinner I try to read the newspaper. Daily newspapers reach Ulipur the next day so I read the old news. At around 9.30 or 10 p.m. I lie on my bed and try to make calculations about what I expected to do in my life and what I am doing now. I think about my family and go to sleep (December, 1997).

My discussion on the personal lives of field workers has become an account of their personal problems. It is particularly striking that field workers work very hard. Some NGOs or centres have more than 90 per cent repayment rates on their credit or good attendance in the nonformal schools. So why bother about their personal problems? I remember that Syed Hashemi asked me the same question (Dhaka, April, 1998). In a sense he is right, since unemployment is very high so NGOs will never have a scarcity of new field workers even if people cannot bear these problems. Neither Northern NGO management nor the People in Aid Code would agree. Personal lives affect professional life and professional lives affect personal life. In the next Chapter I shall discuss the professional lives of the field workers.

Chapter 6

The Professional Lives of Field Workers

Introduction: Beginning Work

Before my discussion on the professional lives of the field workers it is necessary to examine the work culture of my study NGOs. I would align myself with the opinions of Uphoff (1995) that NGOs in Bangladesh are behaving more like business organisations than the Third or Nonprofit Sector. Unlike the staff in the Christian nonprofits in the USA who do not work for money and represent a genuine expression of their organisations' Christian caring for their employees (Jeavons, 1994), field work is a profession in Bangladesh. The productivity of the field workers then becomes an important issue. Although it is difficult to measure the productivity of staff in the service sector, I would say that NGOs in Bangladesh are obsessed with the 'performance' of their staff in delivering services like microcredit, education, health-awareness etc. I have discussed the level of donor-dependency of NGOs in Bangladesh. Donors give funds for certain activities and evaluate the impact of that 'aid' by certain criteria (accessibility to the target population; improvements in education such as dropout rates, enrolment, girls enrolment; repayment of credit etc). So, to ensure regular funds NGOs have to ask their field workers to maintain performance and meet the donors' criteria. This seems natural as NGOs in Bangladesh are not membership organisations. The NGO agenda in Bangladesh is largely donor-driven, NGO's here being, as usual, accountable to their donors, not their clients.

So, what is the work culture in the study NGOs? All but the SCF 'partner' NGOs have service rules and policies on the promotion and transfer of field workers. In the SCF 'partners' small size and more inter-personal relations seem to have created less necessity for these. The number of staff and large area covered are the major reasons for the presence of these formal rules and policies in PROSHIKA and RDRS, while in the case of MCC this is mainly due to the international and missionary nature of that NGO.

All my study NGOs have clear policies on casual leave, medical leave and other benefits, as discussed in Chapter Three. Although my study

NGOs have these policies, two issues deserve mention. *Firstly*, many of these policies have been formulated only in response to many years of demands from the field workers. This has even in some cases meant personal sacrifice, as forced transfer or redundancy may be imposed for even raising these issues. *Secondly*, the mere existence of these policies is not enough. Most field workers know little about these policies. Many asked me 'What can we do when our superiors do not follow the rules? We cannot go on strike or afford to go to court'. This is a key feature of staff management in NGOs in Bangladesh.

Only MCC has effective rules and hours of work. MCC field workers do not have to work at weekends or after office hours. In that sense they are really lucky. For the rest, any hours of work specified on paper are meaningless. Each field worker will have a charter of duties and knows what performance indicators they must reach. This controls their use of time. In theory, a five-day week is worked in Bangladesh, but organisations have discretion whether to close, for example from midday Thursday to midday Saturday. All are required to close on Friday, but for field workers it is easier to find clients at home then, so many work on Fridays.

So, why do field workers work longer hours? The answer is they need to keep their jobs and need to be lucky and qualify for promotions. Staff evaluation is a major activity which directs field workers on what and how much to do to keep their jobs. NGOs know they can place these criteria on their field workers. If any field worker fails to maintain the standard, they have to leave their job and due to high unemployment, NGOs have no difficulty in recruiting new field workers.

Unfortunately this Chapter, like the last, will largely be a litany of the professional problems of the field workers. I shall discuss the problems related to transfer, promotion, group formation, training and workload of the field workers of my study NGOs. I shall describe how field workers cope with local adversities when people think they are working for 'foreign', 'Christian' organisations and involved in 'evangelism'. Also I shall describe how field workers deal with local government administration and the political power structure.

Over the years, NGOs like PROSHIKA and RDRS have grown into large bureaucratic bodies. I shall discuss in this Chapter how field workers assess their status in that structure and their level of job satisfaction. With the rapid increase in the number of women field workers, field work as a profession has developed a new dimension. I shall discuss what men field workers think about their women colleagues and what women field workers think about the behaviour and attitude of their men colleagues towards them.

At the end of the last Chapter I questioned how NGOs are working (some are doing very well) despite their field workers facing so many personal and professional problems. In this Chapter I shall discuss the strengths of the field workers, which I think work as a driving force behind the functioning of the NGOs against all odds. I shall also not hesitate to mention some of the weaknesses of the field workers which require due attention from policy makers and NGO managers.

A question may arise, do the NGO field workers report their problems (both personal and professional) to their superiors? In this Chapter I shall describe the sad story of a field worker which will once again show how crooks in the NGOs in Bangladesh exploit rural youth due to their poverty and unemployment.

Job Issues

Training

The next section will explore why NGO policy makers see training as valuable. The problem of poor quality training or lack of importance placed on the training of NGO field workers have been discussed by several authors (Griffith, 1987; Adair, 1992; Vivian and Maseko, 1994; Ackerly, 1995; Montgomery *et al.*, 1996; Goetz, 1996). My findings support theirs. The importance of training for the professional development of field workers needs no elaboration. Unfortunately, I have heard many complaints from the field workers on training.

I have already mentioned that most field workers were so poorly trained they did not know the exact definition of their target groups. This is totally different from the findings of Garain (1993) who found training of field workers very effective. The major complaints of the field workers against the training systems of the NGOs are:

1. They are not effective or useful for the professional development of the field workers.
2. Some field workers told me that they did not enjoy listening to the discussion of useless issues for hours.
3. Some field workers told me that many training programmes are irrelevant to their activities. They told me 'we work on credit, our job performance is measured on credit. Why should we go for training on health or homestead gardening?'.

4. Many field workers told me that some training programmes overlap and could be given in one series. For example, the training on health education and nutrition have overlapping content. Field workers told me that these could be given in one module.
5. Most field workers told me that even if they want to go for training they prefer to avoid it because absence from the field for weeks affects their work. Sometimes other field workers take over in the absence of a trainee field worker, but due to workloads this is very difficult. Ultimately this absence from the field affects the evaluation of their performance as measured on the basis of loan disbursement and repayment, distribution of latrine slabs or attendance in nonformal schools. So, field workers prefer not to go to training.
6. All field workers told me that a major drawback of the training programmes is that the impact of training on field workers and clients is not followed up.

So, training has become a routine and unpopular activity. Some training programmes are very useful, for example record keeping, project identification and homestead gardening. If there were impact evaluation of the training programmes, field workers could identify the useful ones. Usually field workers prefer a short period of training (one-two weeks) which affects their work less.

There are exceptions too. MCC spends 10 per cent of its budget on staff development. I have found some positive aspects of the training activities of MCC, which I shall discuss below. The huge investment that MCC makes in staff development may be mainly possible due to the missionary nature of this NGO, under which principles are more important than results. The recent tendency of the NGOs to become self-reliant would restrain them from investing adequate money in staff development. For example SCF 'partner' NGOs have no training programmes. SCF (UK) helps them send their field workers to visit other NGOs.

I appreciate MCC's system of focus group discussion with field workers which decides future training programmes. MCC has three types of training:

1. Policy-based training for senior managers.
2. Activity-based (agriculture, health, disaster preparedness) training for field workers and mid-level managers.
3. Practical training such as cleanliness and primary health-care for support staff so that they can keep the office and its equipment clean.

MCC also sends its field workers to visit other NGOs. Some MCC field workers went to visit India for training too. I would repeat here, this is mainly due to the availability of resources and the missionary nature of this international NGO.

PROSHIKA has an impressive training centre situated outside Dhaka. Their field workers gave me some suggestions for improving training programmes. For example:

1. Only suitable field workers should be asked to go for training. This is a major responsibility for mid-level and senior managers.
2. Training should always be followed up and evaluated by the trainees.
3. Training should be more applied and useful for the field workers.
4. Training should be for a short duration and usually for two-three weeks.

One PROSHIKA field worker who worked for the Grameen Bank told me that it has a better and stricter training system. Field workers are taught its accounting policy; working methods, like how to form groups; duties and functions of field workers and group secretaries and data collection on the clients, which helps to evaluate their work. To become a permanent member of Grameen Bank staff, field workers have to pass examinations after training, in addition to demonstrating satisfactory service. He told me PROSHIKA should follow some of the Grameen Bank's training methods. Surely each NGO should try to learn from other NGOs and take the useful methods for themselves.

Workload

The workload of field workers deserves discussion. When I was talking to the senior and mid-level managers of NGOs, all of them told me that, with the present amount of resources, field workers do have to bear the amount of work they have. I have found some interesting opinions on the workload of the field workers. Among the four study NGOs, MCC, PROSHIKA and RDRS field workers described their present workload as heavier than they can bear (compare Wood, 1994 on PROSHIKA). On the other hand, the SCF 'partner' NGO field workers told me that they have a lesser workload after the handover. Recently, the Bangladesh government has introduced a five-day instead of six-day week, with Saturday as well as Friday free of work. Most SCF 'partner' NGO field workers were critical of this decision. They told me that it has compelled them to work longer hours on 'working' days, which affects their domestic responsibilities. When I asked them why they do not like the two-day holiday, they said that they have to work hard to keep their organisation running and their job depends on their hard work. The reason for this opinion is perhaps that they now feel they are working for the NGOs which they have formed, so their NGO and their job depend on them. They are prepared to do hard work. PROSHIKA and RDRS field workers very often work during weekends, apparently to secure their jobs, which depend on their performance in microcredit, education and other programmes.

The problem of workload affects women field workers differently as they cannot work at night. In that sense women field workers are in an awkward situation because they cannot work longer hours even if they wish.

In the case of MCC each field worker has to work with at least seven groups (around 20 members each) and most field workers told me that they could give adequate attention to each group if they had five, or even four. As a rule each PROSHIKA field worker has to work with 60 groups (each with 15-20 members). Most of PROSHIKA's field workers told me that clients could be served better if they had only 50 groups. Each RDRS field worker has to deal with client groups of at least 450 households (maximum one member from each household). Here again, most field workers want it to be a maximum of 400 households.

There are several reasons why field workers want to work with fewer clients and activities. What field workers emphasised to me is that a major problem is the huge number of clients and another problem is that they have to perform several activities but that they cannot give adequate attention to all of them.

There are other problems too. A major one is the paucity of educated clients to maintain group records. This compels the field workers to spend a long time on record keeping which the clients are supposed to do. In this case I would blame the NGO for not training its clients in record-keeping. Another problem mentioned by many field workers (most commonly by MCC field workers) was that they have to do so much paper-work, like reporting, the evaluation of some of which they found irrelevant. All field workers told me that their paper-work helped them to get a clear picture of their work and helped in future planning but that some of it is simply useless.

Another problem which was reported to me was one related to work with older groups. As a rule, MCC groups should be able to run their own affairs independently after five years and RDRS groups after four. Field workers are supposed to give less time to the graduated groups. But due to a lack of expertise in record-keeping, graduated groups remain dependent upon field workers. Also, the quarrelling tendency about leadership and decision-making among the clients compels the field workers to placate people, which takes time. Above all, this highlights a major problem of NGO activity in Bangladesh - the dependent relationship between NGOs and their clients.

Once again, I would mention that mid-level and senior managers do not think that field workers are overburdened with work. They cite their own experience that when they were field workers they could do it, so why not the present field workers? Some senior and mid-level mangers told me that they suggest that their field workers make their workplans in such a way that they can save both time and resources. For example, one RDRS Thana Manager told me that he suggests to his field workers that they inspect tree plantations on the way to group meetings to save time and money.

Promotion

The importance of promotion for field workers for keeping them motivated and getting good services needs no elaboration. I would repeat here that I have heard many complaints on the promotion system in my study NGOs. I take BRAC's case as an exception (Goetz, 1995).

I agree with Kirlels (1990) who found very few women in the higher positions of NGOs. The two major reasons for this situation are a) women's problems in maintaining the dual obligations of home and work which restrain them from giving the efforts required by their NGOs for promotion, and b) cultural constraints on women's authority (Kirlels, 1990; Adair, 1992; Goetz, 1995). I have found the same and I am afraid that very few women are able to overcome those problems and get promoted.

Some statistics are relevant here to elaborate the gloomy picture of the promotion opportunities of NGO field workers. Of all the field workers interviewed with the questionnaire, only 14 per cent got promotion in their present NGO and five per cent did not accept their promotion (all in MCC discussed below). Among the field workers who got promotion, 86 per cent are men. Almost all of these were promoted only once, but one had been promoted three times. When asked whether they think they will have any opportunity for future promotion, only 53 per cent were optimistic.

NGO promotion opportunities at field level seem similar to those in business firms. In the case of business firms, the expansion of activities and resulting promotion opportunities of the staff mainly depend on the profitability of operations. Similarly, expansion and resulting promotions in NGOs largely depend on their 'performance' and availability of funds from their donors. This seems to support the argument made by Uphoff (1995) who   said that NGOs tend to be more business-like than nonprofit. The problem is that when NGOs expand due to the hard work of the field workers the latter do not always benefit. In many cases expansion results in more bureaucratisation at the mid and top level, creating frustration among the field workers.

The major complaints that I heard from the field workers about promotion are that it is not systematic and in some cases is corrupt and full of flaws. Among the study NGOs (except the SCF 'partners') there are formal rules on promotion. The paucity of promotion rules among the SCF 'partner' NGOs do not seem to be unusual considering their size and relatively flat staff structure. I have heard many complaints of corruption in the promotion system. One complaint is that many advertisements for higher posts in NGOs are not put in the newspapers or advertised by internal circulars so that field workers could apply. Some field workers complained to me that sometimes they know about the advertisement at the eleventh hour and they cannot find the time to furnish the necessary requirements for applications. Flaws in the staff evaluation system and complaints about personal 'connections' are very common. Some field workers told me their of astonishment when they find new field workers and superiors during staff meetings, training or in the office without knowing how or when he/she got the appointment. Some NGOs are dominated (to some extent controlled) by one or a few charismatic figures. These people do not bother about the rules. A PROSHIKA Thana Manager who worked for GK told me that GK's Executive Director is the law, and rules are not relevant to him. Of the three study NGOs who have a formal promotion system, promotions are mainly based on interviews and evaluation.

A major discontent among the field workers without a BA/BSc/MA/MSc/MSc/(Agr) degree is that they cannot compete with the more qualified field workers for promotion. They argued that their assets are experience and dedication, which they think deserve appreciation. At the same time, NGOs require managers with some skills in written and spoken English, which makes the opportunity for promotion of non-graduate field workers weaker.

I found a disappointing feature in MCC where field workers are transferred to far away districts after promotion. So no women and not all men field workers accept promotion. Women field workers told me that they cannot move with their families so they want to be posted in Noakhali-Lakshmipur after promotion. They also told me that higher posts in MCC are usually filled with people from outside. Most MCC men field workers who can move after promotion usually take it.

I have found a very clear system of promotion in PROSHIKA. PROSHIKA field workers have to take a two hour written examination and an interview. In promotion, evaluation is given due importance. Field workers told me that the positive aspect of the system is that many good field workers pass easily, but the negative side is that good field workers who are not good academically find it difficult to pass. Above all, to take the interviews and examination field workers have to do some reading, which is difficult with the heavy workload.

Transfer

I heard many complaints about the transfer system, except in the SCF 'partners'. Since SCF 'partner' NGO field workers are local women they think it is convenient for them to work in their own and neighbouring villages. The utilisation of the family backgrounds of the field workers by the SCF (UK) management has already been discussed. All Directors of the SCF 'partner' NGOs told me that clients usually do not complain against the women of their own area, and that if there is any complaint they usually sort it out through discussion.

The three other study NGOs have severe problems with transfer. In general, field workers have to be transferred to a new area or station every three years. But there are many examples where a field worker is transferred twice in a year and some field workers are allowed to work in an area for four to six years. All field workers told me that usually it takes 6-12 months to build trust among the clients in a new working area. So, they think frequent transfer greatly hampers their work. When I was talking to mid-level and senior NGO managers they gave me some opinions on the transfer policy of their NGOs. *Firstly*, there is a human side to transfer. Some field workers are deemed indispensable for more than three years to the management for an area due to his/her sincerity, clients' requests etc. Also, there is a human side where most field workers prefer to work near their home or in places with better living conditions (good schools, electricity, transport etc.). *Secondly*, NGO managers sometimes use transfer as a means of punishment for dishonesty, insubordination, inefficiency etc. The problem arises when an NGO or its managers follow a double standard for different field workers. Many field workers complained to me that some get their suitable posting by just applying but others do not. They told me that standards are not maintained equally for all workers. Personal biases or links with management in many cases create frustration among field workers. I found that all mid-level and senior managers find it difficult to deal with the problem of transfer because using 'connections' is a common feature in Bangladesh society, and NGOs are no exception. One day I was attending a staff meeting at Rajarhat and I found the District Coordinator of RDRS Kurigram warning all field workers against any kind of lobbying for or against transfers; otherwise they would face termination.

PROSHIKA usually allows its field workers who are married couples to work in the same place. Some couples told me that they had to wait years to get a posting in the same place or station. Regarding this complaint, senior and mid-level managers told me that the paucity of posts for couples is a major hurdle.

All mid-level and senior managers supported the three year transfer policy. Some field workers said if they work well they should not even be transferred after five years, because transfer causes problems like changing schools for children, looking for a new house etc. The arguments in favour of a three year transfer system are that field workers become known to everybody in three years and sometimes they delegate some of their work to reliable clients or (in the case of RDRS) volunteers which hampers the smooth functioning of the groups. Some senior and mid-level managers told me that after transfer the new field worker can find out the flaws of the old field workers which helps them to evaluate the activities of the old field worker and rectify them.

Status at Work

Until recently, BRAC's lowest level of village worker, the Programme Assistant, was not part of its regular staff structure. Goetz termed the field level workers as 'kutcha' (raw) bureaucrats, which indicates the contingent, impoverished, ambiguous role of field workers. Most important, however, is the fact that field workers of the NGOs may be in the least desirable positions in their organisations from a career point of view - careers are not made in the field, nor on women's programmes (Goetz, 1996). My findings strongly confirm Goetz's comments.

The interaction between the field workers with his/her clients and superiors will be elaborated in the next Chapter. In this section I shall discuss how the 'kutcha bureaucrats' assess their position in their organisations. All field workers told me that they always remain under pressure by their superiors to fulfil their organisations' targets. When they fail they are treated like criminals. In many cases they cannot explain the reason just because they are not allowed to do so. The immediate causes appear to be: excessive importance on showing performance, poor job security for field workers and the absence of trade unions in NGOs.

Most field workers expressed to me their dismay at the maltreatment they face from their superiors. Field workers are maltreated for being late, or poor in performing their activities. Above all, all of them told me that they do not get their due status in their organisations. Some NGO managers really seem to become too strict on their staff to get their work done. This creates great frustration among the field workers, which could apparently be easily resolved through some minor changes in the attitude and policies of some mid-level and senior managers. For example:

1. Mita Rani, Homestead Development Facilitator, Maizdee, MCC: She told me that when visitors from outside come to visit the clients, a superior always remains with the visitor so that the field workers or clients cannot say anything which might affect the superior's job (March, 1998).
2. Abdul Quyyum, field worker, Kurigram Sadar, RDRS: He thinks field workers are not counted as RDRS staff. He told me that field workers are sometimes abused by their superiors when they come back from the field before 5pm. He also told me that their superiors send night-guards to the local markets to see whether the field workers are chatting in the market. He told me 'we are neither clients nor staff of our NGOs. We are treated like servants' (December, 1997).
3. Abdus Salam, Assistant Thana Manager, Rajarhat, RDRS: When he was a field worker very often his superiors went to find him at his mess at nine at night, whether he was working or sleeping (December, 1997).

All field workers of RDRS and MCC interviewed complained to me that their organisations publish newsletters which they are forced to buy. In Kurigram, I found great discontent among the field workers about having to buy tickets for a raffle which they thought was forced selling by their NGO. I found the Thana Managers were in great trouble because they had to sell a certain number of tickets among the field staff even though they were not willing. Interestingly, SCF 'partner' NGO field workers told me that before handover their superiors forbade them to talk among themselves and that they felt very insulted. Now, since they have formed their own NGOs, they can talk freely, which they think is a great achievement.

External Relationships

Jackson (1997a) in her study of field workers in India found that they faced rejection from villagers, who did not want anyone to come to their villages.

I did not find such strong resistance from the villagers against group formation. Field workers face initial suspicion and resistance in starting work in any village. At least, in Bangladesh the language is always Bangla, so the thing field workers have to do for communication is to learn and understand the local accent.

To build trust among the clients and local community, field workers have to prove their commitment. Most field workers told me that they approach their target population slowly and always keep good relations with local religious and political leaders, teachers etc. Regarding Goetz's (1996) observations on the formation of women's groups, I found the same and I agree with her that the easiest way to get access to a woman is to go to her husband and to entice him with credit. I should say that the only access to Bangladeshi rural women is through their husbands (if married), or parents or elder brothers in the case of unmarried women. To even approach women field workers it is usual in rural Bangladesh to contact the male members of either their family or village,. Other problems faced by field workers like fear of religious conversion or husbands' fear that their wives will be made disobedient are also related to the social structure.

While talking to field workers, I was informed that there are several problems in forming client groups. The major problems are: suspicion from clients, difficulty in finding suitable clients as per the criteria of the NGOs, competition among the NGOs for clients, resistance from local influentials and religious leaders, clients' lack of time and the restrictive rules of the NGOs (like requirements for savings, attendance at group meetings etc). Above all, the eagerness of the clients to reap quick financial and material benefits works as a major problem in group formation. These problems are elaborated below.

1. *Suspicion*: Suspicion from the clients is a major problem for the field workers. When field workers approach their clients they face suspicious attitudes and comments like 'field workers will flee with our saved money'. The major reason behind this kind of suspicion is that there are many fake NGOs in rural areas who have cheated their innocent clients. Also, there are many examples of cooperatives formed by some crooks for embezzling the money. Many field workers told me when they tried to approach their clients many of them did not even like to talk, to come out of their house to talk or even to offer them a seat. Another problem for the field workers is that they are suspected of being involved in Christian evangelism, a frequent suspicion in Bangladesh. One Hindu Homestead Development Facilitator of MCC told me that when she started to try to form groups most people told her

'you are a Hindu so you have come to work for the Christians to convert us into Christians'. Most NGOs are suspected of evangelism and cheating. The problem is more severe for international NGOs. For example:

> *Masuma, field worker, SCF 'partner', Naria: Masuma joined SCF (UK) in 1991 and her superiors told her to form at least three groups in two months. Initially she faced immense problems in forming groups. One day, she saw some people were making fishing nets on the bank of the river. She approached one but he told her that he is suspicious that Christians will get their loans back at any cost. She said that she was not a Christian, she was from his village. Then the man advised her to approach his uncle and convince him. The uncle talked to the man, who agreed to allow his wife to become a SCF (UK) group member. Other women from that village followed her. In a neighbouring village she tried to form groups and faced the same problems. A women from the second village said that she would be a member if the field worker could arrange a loan to pay for the fees for her son's SSC examination. The field worker agreed to do that and managed to form a group of seven women. Two months later, the field worker gave the money to the woman for her son's examination. The next morning the borrower woman came back with the money saying that everybody in her village had threatened her because it is not permitted for a Muslim to take money from Christians, who will take back the money with interest at any cost. The field worker assured the woman that it would never happen and she had nothing to worry about. The son of that woman passed the SSC and is now working in Bahrain* (November, 1997).

2. *Resistance from religious leaders and influentials*: Due to purdah, this problem becomes more acute when men field workers have to form women's groups. One field worker told me that a religious leader told him that it is a sin for women to talk to unrelated men. The field worker replied 'what is more important, not to commit a sin or earning a living when you have no food?' The way SCF (UK) approached the local influentials and recruited their daughters has already been discussed. Resistance from the local influentials is not surprising. Powerful men do not like to see poor people getting organised and achieving economic and social lift. Most field workers told me that to form groups in any village, they approach a local Union Parishad member or

chairman and get his permission to work in the village. Sometimes they have to attend the meetings of the local Union Parishad (where all men gather) and approach their clients in front of the members.

3. *Time and resource constraints*: These are the two major constraints which discourage many clients from becoming group members and attending meetings regularly. Time is a major constraint and men clients usually prefer to meet in the evening after work, which restricts many women field workers from working with them. In that sense, women clients are easier to approach, but many field workers told me that their constraints are different - inability to leave children at home, poultry unattended etc. Also there are certain times of the year when women become very busy with the processing of crops, especially the milling of rice, which leaves little or no time to attend meetings. Many poor women work during these times and do not like to forgo their income. Many field workers told me that the rule of saving a certain amount of money in the group fund regularly prevents many poor men and women from becoming NGO members.

4. *Availability of suitable clients*: Field workers always try to work with clients who will attend the group meetings, repay loans regularly and co-operate with them. A major problem in some areas of Bangladesh is river erosion which destroys many people's home and land in a matter of hours. Many field workers told me that it is difficult to work with clients whose livelihoods are vulnerable to the might of the river. Displaced clients are difficult (in some cases impossible) to locate, let alone contact. Some poor men migrate during sowing and harvesting seasons in search of work; they are another difficult group to work with. Even if suitable people are found, then comes the problem of mutual trust among the clients. Due to the presence of many economic and social conflicts, some clients object to the presence of one or more members of his/her group saying 'if he/she becomes member I will not be a member'.

5. *Eagerness to get financial or material benefits*: All field workers told me that people think that if they become NGO group members they will get money or material relief. The relief operations by many agencies after natural disasters has created this mentality. Before becoming members many people ask which benefits they will get and how soon. Field workers say that in many areas people can now choose among NGOs. This disreputable competition among the NGOs is a real danger for 'development' at the grass-root level. Many people think NGO group membership will enable them to get free seeds, poultry, medicine etc. and demand it as a condition of membership. Many field

workers told me that it is difficult to work in this situation. The recent fashion for giving microcredit has created another competition among the NGOs since members leave NGOs on the pretext that the other NGOs give bigger more loans quickly. I have found this problem to be acute in some RDRS working areas.

Local Elites

There are many examples of field level bureaucrats and political leaders becoming antagonistic to the NGO field workers (compare Griffith, 1987; Sen, 1999 on India). The Area Coordinator of PROSHIKA Sakhipur told me that when he was working at Agoulzhora Thana of the Barisal District, the workers of the ruling political party took away his and his colleagues' motorcycles. He went to the local leader of that political party and asked for his help, but failed. He then went to the district level leaders of the party and told them that if he did not get back the motorcycles in 24 hours he would make a formal complaint to the national level leaders of the party and to journalists, some of whom he knew. He also told them that if he reported to the journalists this would tarnish the image of the party. After several meeting with the party leaders, the situation improved gradually and the workers returned the (damaged) motorcycles and apologised to him. The same AC of Sakhipur told me that for the inauguration of their new office at Sakhipur he invited the TNO[1] of Sakhipur. The Executive Director and some other senior managers of PROSHIKA were also invited. After the function there was a tea party arranged in a small room. Due to space constraints, the TNO could not be accommodated with the Executive Director and other senior managers in the same room. Instead, the TNO was entertained in another room, with PROSHIKA field workers. The TNO felt very dishonoured and when the AC went to invite him to another PROSHIKA function the TNO rejected the invitation straightway even though the AC had apologised for the previous incident.

In many cases government bureaucrats do not like the popularity and influence of the NGOs. NGOs like PROSHIKA, RDRS and MCC have better logistic facilities and attractive offices and equipment like computers, which government offices and their staff lack. Above all, bureaucrats traditionally enjoy immense power at the local and national level in Bangladesh.

There are other problems in dealing with government staff. Most women field workers complained to me that very often women clients ask

---

[1] TNO (Thana Executive Officer) is the top government bureaucrat at the thana level.

for contraceptives, which they cannot give. They said that the problem with government family planning workers is that instead of visiting the villages for which they are responsible they stay in their offices or towns. Some women field workers told me that government field workers, like health, family planning and agricultural extension workers, become very angry when NGO field workers ask them to go to their clients or to look after them saying 'I will do my work, you don't have to tell me!'. Some NGO field workers told me that government staff sometimes feel jealous of them due to the misconception that NGO field workers enjoy better salaries, benefits and logistic facilities. Some NGO field workers told me that government field workers do not come to them unless they are in trouble. One RDRS man field worker told me that just two days before an immunisation day, a government health worker came to him for help in bringing a huge number of children for immunisation because his RDRS supervisors might come for a surprise visit and if they found only a few children for immunisation the government health worker's job would be at stake. One MCC field worker told me that he finds it difficult to meet the government agricultural extension worker. He also told me that one day he was surprised when the government extension worker showed his superiors the high quality vegetables of an MCC client as his own success.

All this lack of co-operation and suspicion from government field staff to their NGO counterparts is familiar. For decades, 'development' has been the responsibility of the state and due to the long colonial legacy and paucity of democracy, the state machinery has become so powerful, inefficient and disconnected from the people that in many cases it cannot accept the successes of the NGOs and becomes reluctant to co-operate.

Local political leaders know the limitations of NGO field workers, that they cannot get into a confrontation with them. Antagonism from the local political power structure makes the work of the NGO field workers very difficult or impossible. Field workers and mid-level managers are really in a critical condition in these power plays at the local level. Dealing with the antagonistic bureaucrats and political leaders is a very sensitive issue.

Gender

Women Talk about Men

Rao and Kelleher (1995) report on the criticisms women field workers face from their men colleagues of being weak and inefficient which are mainly due to the men's patriarchal attitude. They also reported some success in ameliorating these problems through the Gender Quality Action Learning Programme in BRAC (Rao and Kelleher, 1998, sections 4.1.3 and 4.5). Goetz (1995) found that women field workers are rarely given the opportunity to prove themselves. Also women may withhold their views and solutions because they know these will be accorded less value than those of men. As a result, they are also accused of lacking initiative, and competence and would be forced to keep quiet (Goetz, 1995; Daniels, 1988). I found the same in my field work.

Adair (1992) also noted that the management of PROSHIKA made a decision to ensure the safety of women staff. The management of all study NGOs have this policy and I have found that most mid-level and senior managers try to maintain it. At the same time, women field workers get told (directly or indirectly) that they are enjoying privileges which men field workers cannot demand. For example, sometimes mid-level managers help the women field workers in their work, ask the men field workers to help their women colleagues or do not ask the women field workers to do risky or tedious work etc.

I have heard several other problems from the women staff on the attitude of their men colleagues towards them. All women field workers think that they are better than men because they are more obedient and sincere, although their men colleagues do not accept this. One woman told me that a woman faces double criticism from men; for inefficiency and for being women. All women field workers told me that when NGOs target women, it is much more feasible to work through women staff than men.

Most women told me that their men colleagues underestimate their cost of living and work. One told me that a rickshawpuller will charge a woman passenger more. Due to purdah, women field workers cannot travel freely in public transport like men. Sometimes drivers or contractors of public transport do not like to take women passengers because they have no reserved seats (in Bangladesh there are reserved seats for women on public transport). One women field worker asked me how could her men colleagues tell her to stay in the office or in the field after office hours? Would they allow their wives to do so? This is a strong argument.

One woman field worker told me that women field workers are more expensive than men because they are paid three months salary during maternity leave and they cannot work long hours like men. She thinks her NGO uses her to get access to women and good repayment of microcredit. She also told me that her NGO shows the donors that it is recruiting women. She thinks her NGO benefits from her hard work, but she does not. Rina Sen Gupta gave me a similar opinion. Rina started her career as a field worker. She thinks her work position has changed because she became the

Regional Director of an International NGO, but her condition as a woman, wife, mother in her family and society has not changed (Dhaka, April, 1998). I find these opinions very interesting and useful in understanding how little the status of women in Bangladeshi society has changed.

Men Talk about Women

I agree with Rao and Kelleher (1995) that men field workers devalue women's work and try to exclude them from decision-making. They argued that women should be men to succeed in NGOs (Rao and Kelleher, 1995). Such talk is common in rural Bangladesh.

I heard some interesting opinions from men field workers on their women colleagues. *Firstly,* many men told me that NGOs are recruiting women because of donor pressure or to please their donors. *Secondly,* however, most men told me that there are both good and bad women field workers so they do not like to generalise about all women field workers. *Thirdly,* some men field workers complained to me that many women field workers cannot work independently and very often depend on the help and assistance of their men colleagues and superiors. I have found a little evidence for this argument. *Fourthly,* some men told me that women field workers' jobs should be restricted to dealing with cash in the office or in teaching. This argument seems unacceptable when there are many good women field workers. *Fifth,* most field workers and clients told me that women clients find it more convenient to deal with women field workers. One RDRS man field worker, while giving me his opinion on women field workers, told me the positive aspects of recruiting women field workers are that it has helped to reduce the unemployment problem of women and helped women field workers to become self-reliant. He mentioned the negative aspects too; that women field workers cannot work after dark, depend on help from men colleagues or superiors and face more resistance from the fundamentalists and religious leaders than men.

Women's Problems

Most of the professional problems of women field workers have already been discussed. In this section I shall mainly deal with the issue of why we find very few women in the mid or senior level management of NGOs. Most women field workers told me that the nature of their work demands huge stress and long working hours, which, with their domestic responsibilities and social and biological constraints, they cannot perform better than their men colleagues. Many field workers told me, since credit repayment is a major indicator of job performance it is easier for men field workers to force their clients than women. There are some extraordinary women field workers who are really sincere and dedicated and some do get promoted. But all these successful women lamented to me the sacrifice they had to make of their families, particularly children. One woman Thana Coordinator of PROSHIKA, who was promoted from field worker to that position told me 'you have to become a man to become a successful woman field worker'. It seems that most women field workers are constrained from becoming professionally successful, even if they want to be.

Evaluation

Weaknesses of the Field Workers

Griffith (1987) identified problems with the field staff of NGOs, which I also found. Griffith found that field workers are concerned with satisfying their superiors and only approach the easily accessible clients. These complaints seem unfair when NGO management has become too 'success' or 'performance' oriented - microcredit is a good example. Still, my research suggests that the strengths of the field workers are much more prominent than their weaknesses, which is a major reason for the success of the NGOs in Bangladesh. The weaknesses that I shall discuss below could be easily overcome by better training and supervision and the commitment of the higher and mid-level management of the NGOs. For example, poor record-keeping by the field workers could be easily overcome by training and close supervision by superiors.

Some weaknesses of the field workers have already been discussed. Some other weaknesses that I have observed are poor supervision of group activities, poor planning and in a few cases corruption. Due to insincerity or lack of time some field workers do not carry out their activities. I have heard some complaints on poor supervision, such as tree plantation programmes where field workers disbursed wheat in payment to the caretakers but the trees were not properly cared for. In some cases field workers fail to identify or mobilise their clients to guard against the entry of shrewd people who become group members to create trouble, or then to mobilise other members to discipline those criminals. When field workers mobilise their clients to take decisions by themselves, groups work well. Unfortunately, some field workers fail to understand the importance of this, and groups become dependent.

I have also heard a few allegations of corruption by field workers, which may have been due to laxity of supervision by managers. Similarly, the problem of poor planning by field workers could be easily overcome by training and supervision. A good example of poor planning by some field workers is fixing the date of group meetings on the days of periodic markets. In rural Bangladesh, many markets are periodic (once or twice a week) and clients and their families are busy on those days. Attendance at group meetings is very poor on market days. Also, to make clients punctual, field workers should be punctual. I have heard and witnessed many instances of lack of punctuality by the field workers, which could be easily avoided.

A major weakness of many field workers is their inability to maintain strict discipline and impose it on their clients. I witnessed a major problem in RDRS where a client managed to join two groups simultaneously and applied for loans from both groups. This means major laxity by the field workers. I have also witnessed poor attendance in nonformal and adult literacy classes when field workers fail to maintain discipline. Here again, close supervision by mid-level managers could play a key role.

Some field workers seemed to lack honesty, which deserves real attention from the mid-level and senior managers. For example, when I was working in the RDRS study area, RDRS was distributing treadle pumps to its clients at a subsidised rate. Some clients told me that they cannot afford Tk 650 for a pump whereas actually they were supposed to pay only Tk 254, the rest being paid by RDRS. This is an example of how dishonesty of the field workers affects the activities and credibility of their NGOs. Field workers should be directed to disseminate the right information, and this is very important where the general tendency among the clients is to perceive the services from the NGOs as relief which needs to repayment. In one PROSHIKA nonformal school a teacher used to beat her students regularly, but the relevant field worker did not discipline her. After one month, the parents of the children complained to the Thana Coordinator who immediately met with the field worker concerned and the teacher and formally reported it to head office. This is another example of the dishonesty of field workers. I blame the field worker for not noticing the problem in time and not taking necessary action.

Some women clients told me that sometimes they or their husbands dislike the eagerness of some women field workers for glamour. Here, training and close supervision by the mid-level managers could play a key role. Goetz (1995, 1996) identified a major weakness among the women field workers. She found that many women field workers were reluctant to identify the similarities between themselves and women clients. She said this might be caused by the low organisational status of the women field workers and by class differences. She also warns us not to assume that women are always more effective than men in reaching rural women. Few people can cross class differences (even in research). This is mainly due to the education system in the country, which is elitist and creates distance between the educated and the non-educated, rich and the poor, working women and non-working women. NGO field workers were supposed to be motivated to interact with their clients very closely while working in the same stratified society where education, income and profession create differences that are difficult (more precisely impossible) to overcome. I can confirm that I, like the field workers, found it impossible. But like most women clients I would prefer women field workers to work with women clients.

Strengths of the Field Workers

Griffith (1987) found field workers in a project in India were suffering from lack of motivation to use their initiative (Griffith, 1987). I would say there are several reasons why field workers lack motivation and shall discuss some of them below. I also think that the management of NGOs very often discourages field workers from using their full potential.

I firmly believe that the strengths of the field workers deserve proper attention and elaboration. In spite of all the personal and professional problems it is the strengths of the field workers which ensure the smooth functioning of the NGOs in achieving national and international acclaim. The strengths of the field workers are invaluable assets for the NGOs and a country, assets which require due utilisation and credit.

Goetz (1995) found that women field workers are more self-critical, responsible, more aware of problems of women clients etc; I found the same in most cases during my research. Heyns (1996) found in South Africa that field workers cannot rely on the syllabus learned on training courses. They have to be able to respond to whatever a situation throws up. Clark (1991) notes that field knowledge is based on eyewitness assessment, and is difficult to 'quantify and tabulate'. It is not stored on paper with charts and tables as favoured by stereotypical headquarter bureaucrats but in the programme staff's own memory. I have found that this important issue is sometimes ignored by senior NGO managers.

I found that some field workers have tremendous power for motivating clients to form groups and providing services like health, education etc. Although my study NGOs (except MCC) are very much pre-occupied with credit, some field workers feel that their credit programmes cannot be successful without other services like health awareness, education etc. Some field workers told me that they motivate their clients by saying that if they did not have a peat latrine they would get diseases, have to spend money on medicine and not be able to work. Then how would they repay the loans? Some field workers proudly told me that although their organisations give top priority to credit they themselves know how to deliver other services which they feel are equally or more important. To motivate their clients for training, field workers told me that they tell their clients that money or assets can be stolen but your skill or knowledge cannot. So, have some training and it is a life-long investment. These are a few examples of how field workers use their wits in helping their clients.

A major duty of the field worker is to keep good contact with his/her clients and to motivate them. All field workers told me that they have to follow different methods of motivating or mobilising their clients, as they cannot force them. For this, field workers adopt different means. One field worker told me that he does not refuse a betel leaf from the dirty hands of his clients because they will feel dishonoured. Another field worker told me that he joined his clients in harvesting the paddy (rice) when the client told him about his inability to hire labourers. These are all strengths of the field workers which have remained unrecognised by most NGO managers and researchers.

I have found many field workers to have intelligence in field work which could not always be acquired through academic degrees. One field worker told me how he convinced his clients to vaccinate their poultry. His clients were not interested in vaccinating their poultry due to cost, superstition etc. He told them 'do you want to lose your Tk 150 valued chicken for Tk 2?'. His clients agreed to vaccinate their poultry. Then clients raised the question of how to vaccinate their poultry when the government vaccinator did not like to come to their villages. So the field worker invented a new idea. He asked all clients to contribute Tk 3 per bird to bear the cost of vaccination (which is actually free), travel costs and 'tea money' for the vaccinator. The government vaccinator came to the village for a whole day and vaccinated the birds of many clients and non-clients for Tk 3 per bird.

How intelligently a field worker can solve a problem is illustrated by an example below:

Abdul Quyyum, field worker, Kurigram Sadar, RDRS: One of his clients, a village doctor, did not repay loans for more than three months. Other group members tried to convince the doctor but he was a very shrewd man and did not pay heed to their requests. Mr. Quyyum told the doctor that he had a good income and he should repay the loans. Then the doctor started to pay Tk 50 per week while he was supposed to pay Tk 200 per week. After paying instalments for six weeks he stopped. The field worker started to visit the house of the doctor every week. The wife of the doctor offered him tea and the field worker told her that he was like her younger brother and asked whether she would like her younger brother to lose his job for her husband. The doctor's wife became very sympathetic to him and started to repay the loan (December, 1997).

Many field workers told me proudly that they feel happy they can bring changes to the lives of many poor. One SCF 'partner' NGO field worker told me proudly that around 60 per cent of shops in the local market are built from her loans, so she and her family always enjoyed special privileges from these businessmen. One PROSHIKA field worker told me that one day he was surprised to see that a primary school teacher had brought his daughter to enrol her in his nonformal school because the teacher thought nonformal schools were better than his school. Some field workers become role models for their clients and others through their activities. One PROSHIKA field worker showed me that with his experience he started a nursery which will give him a profit of around 50 thousand taka in a year. He had also grown a teak garden of 300 trees on one acre of land and a jack fruit garden of 200 trees on three acres of land. He cultivated water melon instead of a paddy on his agricultural land and made more profit. He invited his clients to see his garden, nursery and field to show how money could be made from limited resources. Many field workers told me that they advise their clients not to be disheartened by their limited land and resources. They advise their clients to grow vegetables and plant fruit-trees on the land beside their house or to rear chickens, none of which activities require investments. Field workers told me that some clients take their advice and gratefully acknowledge it.

A major achievement of the field workers is their excellent public relations. Actually, their job demands very good public relations. To do their work, field workers need not only to keep good relations with their clients and their families, they have to keep warm relations with the local religious and political leaders, bureaucrats, teachers and influentials. Also, they have to keep good contact with the staff of other NGOs. Some strengths of the field workers are visible when someone like me travels with them. Some good field workers form such good relationships with their clients and people in their working area that when they visit their former working area they have to stop frequently to exchange greetings. I found much evidence of this when travelling with the field workers. Because of their sincerity many field workers are given hearty farewells on their departure. Some field workers told me they do not worry about getting any help from people in their former working areas because of their popularity. All these good relations have many advantages. A good field worker can ask for help from any of them in time of need. One field worker told me that they have no money like businessmen, goons like politicians, power like the bureaucrats or force like the police but their asset is their human relations with other people. One field worker proudly told me that if he wants to start any business in his former working areas he will not need any capital, his reputation and trust are his capital. Some field workers proudly told me that if they ask their clients to vote for a certain candidate in the local or national elections most of their clients would do so, but usually they do not ask. But I have heard some complaints against PROSHIKA for taking sides in elections; see Hoque and Siddiquee, 1998; Hashemi and Hassan, 1999. Still, during elections many candidates approach the field workers and mid-level managers for favour or support which has made the influence and popularity of NGOs and their workers a problem by bringing them into party politics.

The influence and popularity of the field workers of NGOs have compelled some government and semi-government bodies to invite them to their functions and other activities, which I shall elaborate below. The District Coordinator of Kurigram told me that in every winter or flood the Deputy Commissioner[2] asks him when RDRS will begin its relief work. He said most Deputy Commissioners plainly confess the limitations of the government machineries and ask for help from NGOs like RDRS. He also told me that in one meeting the Deputy Commissioner complained that he did not find any government doctor or health worker during an outbreak of diarrhoea but found many NGO workers working round the clock to save lives.

Another important virtue of the field workers which really impressed me is their mutual co-operation and trust. With a very few exceptions, all field workers co-operate with each other and extend their hands to their colleagues in times of distress. Usually one field worker does not turn down a request from his or her colleagues for help with work or personal problems. This may be mainly due to the good field-level management of NGOs and the commitment of their staff. It is not unusual to find all men field workers taking tea in one restaurant or going to the cinema. This mutual understanding and team-work are vital strengths of the NGOs. I

---

[2] The Deputy Commissioner is the top government bureaucrat at the district level.

have found this cordial relationship among field workers from small 'partner' NGOs to the large NGO, PROSHIKA.

A major strength of the field workers is their innovativeness and power to mobilise their clients. All field workers told me that they have to be innovative in solving their problems. These problems range from non-repayment of credit to absence from group-meetings. The problem of microcredit will be discussed in Chapter Eight. Field workers use their intelligence in solving the problems of absence from group meetings and adult literacy or social awareness classes. Some field workers told me that they simply say 'I have come a long way why shouldn't you?'. Another way of solving this problem of non-participation is mobilising the clients to decide the punishment for it. Some field workers told me that they leave it to their clients to decide on the mode of punishment for non-participation and it works very well. For example, sometimes clients decide to fine an absentee client Tk 5 or 10 which he/she would spend in feeding other members betel-leaf, biscuits or sweets. Another punishment is to count all the money collected at the next meeting (which is really laborious work) or to present the next lesson in the class. Some field workers told me that sometimes they give loans for some purposes which their organisation would never support. For example, if a client asks for a loan for his/her daughter's marriage, medical treatment or house-repairs many field workers consult about it with other group members. They might sanction it, but show in official documents that the loan was given for business. All field workers told me that although risky, these vital decisions positively affect the image of the NGOs since clients and other villagers remain grateful for the generosity of the field worker and his/her NGO at a time of need. These are a few ways in which field workers manage some common problems.

A major duty of all field workers is to fight resistance, and in some cases attacks, from local religious leaders and fundamentalists. A major way out for field workers is to convince these leaders that they are not working against religious values or engaged in Christian evangelism. I have found that MCC, RDRS and SCF (UK) field workers faced more resistance than PROSHIKA because of the foreign origin of these NGOs. To solve these problems, field workers and mid-level managers adopt several measures. They told me that they perform prayers in their office compound and regularly go to local mosques to offer prayers, to break the misconception that they are anti-religion. Some field workers and mid-level managers told me that they keep the gates of their offices open to make it clear to people that their work is 'development', not religious, and they are not involved in any conspiracy or motivation against the religious sentiments of the people. The way field workers face the fundamentalists can be understood from the experience of one field worker.

Abdur Rahim, Thana Coordinator, Sakhipur, PROSHIKA: When he was working in Char Borhanuddin Thana of the Bhola District he faced strong resistance from the fundamentalists and religious leaders who believed that PROSHIKA was teaching anti-Islamic values and ideas in its nonformal schools. The fundamentalists organised a procession against him and PROSHIKA and submitted a memorandum to the local TNO (Thana Executive Officer). Mr. Rahim initially became very afraid and talked about it with his colleagues and superiors. He took all the books from his nonformal schools to the TNO and asked for his help. The TNO checked all the books and said there was nothing objectionable in them. Later, the TNO became very sympathetic to him and his colleagues and helped him on many occasions. One day, one of his students died and he went to her house to express his condolences, but he was not allowed to enter the house. Next day, when he went to the school with the teacher, students told him that the fundamentalists and religious leaders had spread rumours in the village that PROSHIKA people bury the dead in black cloth and do not perform the customary prayers before burial (the tradition is to bury the dead in white cloth and perform a special prayer). Mr. Rahim went to the grave of the student and offered prayers for her to break the misunderstanding. On another occasion the river Meghna started erosion and many people lost their homes and livelihood. The religious leaders started saying that it was happening due to the presence of NGOs in their villages. Mr. Rahim told me that he found it very difficult to deal with the fundamentalists. When he left Char Borhanuddin in 1998 after working there for four years, he told me that the situation had improved a lot (February, 1998).

There are many examples which I can give to show the strengths of the field workers. One SCF 'partner' NGO field worker told me that before handover SCF (UK) decided to sink some deep and shallow tube-wells. Many people, including the local leaders, tried to influence her to sink the tubewells in places convenient to them. They thought that since she was a woman it would be easy to persuade her to get their work done. She asked her clients to vote secretly on the location and then asked the clients to decide on the basis of the results of the votes (compare Mirvis, 1996).

Sometimes when women field workers were explaining their grievances to me on their hard work and problems with child-care and domestic responsibilities I asked them why they were still doing the job. An answer from a woman field worker of MCC deserves mention because it echoes the voices of most of the women field workers:

Instead of doing cooking, gossiping and listening to the harsh words from my husband and mother-in-law it is much better to work for other women and earn some money. (Rokeya Akhter, WDC, MCC, Companyganz, March, 1998).

Case Study

Now I shall describe the story of an NGO field worker which will show how many small and medium sized NGOs in Bangladesh exploit innocent rural youth. This story will also show how field workers face their professional problems when they have no alternative. It will also give evidence of some crooks in Bangladesh who have taken NGOs as their business.

Mohammed Rahim, Development Worker, PROSHIKA. Mr. Rahim's father was a farmer. Unfortunately, his father died of old age complications when I was working at Sakhipur. Mr. Rahim's mother died when he was five. He passed the SSC in the first division in 1982. After taking his SSC he had to stop his education due to financial constraints. He left his village and went to Madhupur Thana town and was admitted into the local college. To support himself and his education he had to live in lodgings with a family and do private tuition. He passed the HSC in 1986 in the second division. He thinks he could not get the first division because he could not afford to have private tuition. When he was studying at college for the HSC examination his father became partially paralysed and the financial problems of his family got worse. His father started to sell his land to support his treatment and his family.

After taking his HSC Mr. Rahim's family and relations advised him to marry into an affluent family which could support his education. He refused to marry young and to become indebted to another family. Rather, he was admitted for a BA degree at Madhupur College. He used to lodge with a family and passed his BA in the second class in 1988. After taking the BA he started to apply for jobs in government, schools, nationalised commercial banks etc. He remembers that he had faced at least 22 interviews and spent around Tk 10,000 as bribes. He failed to get any job and his situation became desperate. One day, one of his cousins told him that a local NGO, SDS (Social Development Society), was recruiting field workers. His cousin also told him that he had to give three months free service to that NGO and deposit Tk 1200 (non-refundable). After joining he would get a food allowance for three months and after that he would get a salary of Tk 520 per month. Mr. Rahim went to Tangail (a district town) to meet the managers of SDS. His ailing, impoverished father could not afford to give him any money. Mr. Rahim mortgaged out some of his father's agricultural land for Tk 2000. Mr. Rahim and three of his cousins from his village joined SDS simultaneously as field workers in 1989. They lived in the SDS field workers' mess in Tangail. Within a few days other field workers told them about the corruption in SDS and Mr. Rahim and his cousins started crying. They met the Executive Director (ED) of SDS and told him that if they did not get any food allowance for three months (which they were promised) how would they survive? After a long discussion the ED agreed to give them Tk 500 per month, which would be deducted from their salaries later. Although Mr. Rahim and his cousins were supposed to give free service for three months, actually he had to give it for five months. After working for one month in Tangail Mr. Rahim was transferred to the Gopalpur Thana office of SDS.

By the end of 1989 Mr. Rahim was promoted to Assistant Field Officer. While working at Gopalpur one day his immediate superior told him and his cousins they would not get their salaries from the following month because their credit repayment rate had fallen below 80 per cent. They asked their superior 'if clients do not repay loans what can we do?'. Their superior replied that he did not want to hear any excuses. Mr. Rahim and his cousins did not get any salary for two months because of low repayment. Mr. Rahim and his cousins went to the clients and told them about their grievances and clients started to repay some money. After this incident Mr. Rahim and his cousins were transferred to four different offices so that they could not band together.

In April 1990, Mr. Rahim married one of his cousins. At his marriage his wife was studying in class eight. He formally took her to his home when she passed the SSC.

In mid-1990, when Mr. Rahim was working at Gopalpur, he received a letter from the head office of his NGO saying that he had been made a member of a nine-man committee to audit the financial accounts of SDS's Sherpur office. As per the rules of the government, NGOs are audited by registered chartered accountancy firms. Later, Mr. Rahim discovered that his NGO actually compelled him to work for the chartered accountancy firm to pay less for the audit.

On his way back to Gopalpur, Mr. Rahim went to Tangail to meet his ED, who ordered him to go to Ghoraghat in Ghatail Thana for two weeks. He did not dare to ask his ED the reason for his trip. On the same evening, he left for Ghoraghat. He was surprised to see spades, shovels and other construction materials in the office compound. He asked the caretaker about it, who did not reply. From the next morning he and eight of his colleagues had to work like day labourers to build a training centre for SDS. He and all his men colleagues had to cut earth and bamboo and carry them on their heads or shoulders. Mr. Rahim's shoulder became swollen so he had to put on straw to reduce the pain. The caretaker cooked for them and the cost of the food was deducted from their salary. Their work was supervised either by the ED or the Financial Secretary of SDS. Mr. Rahim became really ill after working for eight days. He told his superiors about his grievances. He was allowed to return to Tangail and received Tk 300 for the audit and construction work. He was getting a monthly salary of Tk 900. Mr. Rahim and his colleagues later came to know that SDS got funds for building a training centre, but the ED misappropriated the money. When donors wanted to visit the centre the ED constructed it overnight.

During the Eid holiday Mr. Rahim went to visit his family at his village home. His wife and his parents-in-law (who were at the same time his uncle and aunt as his wife as a cousin) became worried at his deteriorating health. He did not tell them about his physical labour because they would not allow him to go back to work. Still, his wife and in-laws asked him not to go back to work, but he did not listen to their request for two reasons. *Firstly*, he would not get any other job and whatever he earned he could at least support himself. *Secondly*, if he had left the job the only occupation left for him was to do private tuition, which would not give him more income that he was getting from SDS.

At the end of 1990, Mr. Rahim got another promotion and his salary was increased to Tk 1200 per month. He was transferred to the head office of SDS at Tangail. There he came to know that SDS was only registered to work in the Tangail district and the district administration complained against its working outside the district. The ED of SDS bribed the District Coordinator (DC) to ignore the situation.

When Mr. Rahim was working in the Tangail head office he had to work in its soap factory. For this he had to cut earth and carry heavy cement bags on his shoulders. His shoulders became swollen again. At that time, a local daily paper reported about the forgery of around 100 thousand taka by the ED of SDS. All the staff of SDS panicked about their jobs and the fear of being arrested. The ED of SDS called all his staff (around 300) and told them politely that if the police arrested him they were to go on hunger strike in front of the local press club demanding his release. The ED told them that he had given them employment and they (the staff) would starve without their jobs. So, the ED told all his staff go to the DC and say that they would starve if he did not release the ED.

It was discovered in a government inquiry that SDS was running a savings programme for children aged between 1-12. SDS took Tk 20 as a non-refundable admission fee for each child and the parents had to save at least Tk 5 per month. The condition was that every member would get 25 per cent profit from their savings after 5 years. Actually, the ED embezzled the money and members complained about it to the local administration and press. The ED repaid most of the money to SDS members and settled the issue after prolonged negotiations. After the settlement of the savings problem, the ED called all his staff and told them that he had no money to pay their salaries. After a long discussion, all staff asked the ED to sell the soap factory and buy some rickshaws and vans with the money. SDS bought around 50 rickshaws and vans and staff started driving them to keep their NGO running. Mr. Rahim could not bear the physical work of driving rickshaws and left Tangail without permission. One day, in mid-1990, SDS authorities sent a letter to Mr. Rahim to ask for the reasons for his unauthorised absence from his duty, and finally sacked him. His cousins from his village continued to work in SDS, but resigned en masse in 1992 after another feud with the authorities.

One of his cousins is now engaged in farming in his village, another is involved in business and the third is working for a local NGO.

After the job in SDS, Mr. Rahim started doing private tuition in his village and some farming. He also started working as a part-time teacher in his own primary school. Towards the end of 1992, one of his uncles ran for the chairmanship in his local Union Parishad (UP) elections and Mr. Rahim campaigned for him. After the elections, Mr. Rahim was appointed as the acting Secretary to the UP. To become permanent Secretary he needed approval from the DC. Mr. Rahim gave one thousand taka to his uncle to expedite his case, but he started dilly dallying. Finally, Mr. Rahim's uncle accompanied him to the DC. But to their disappointment, the DC said that the appointment of secretary should be done through open advertisement. Mr. Rahim knew that he would not get the job. He could not get a job at the primary school either. His uncle gradually shied away from him.

Mr. Rahim joined a local NGO named PDP (People's Development Programme) in March 1994. He joined as a Field Officer and was later promoted to Area Coordinator. His salary was Tk 1400 per month and his work was mainly credit. He had no transport so he had to walk miles every day. A PROSHIKA field worker who was working in his village advised him to apply to PROSHIKA. When he got his job in PROSHIKA, PDP gave him a certificate and released him happily. He joined PROSHIKA as a Development Worker in December 1994 at Char Borhanuddin Thana of the Bhola District.

The story of Mr. Rahim raises several questions, most of which have already been discussed elsewhere in this book. *Firstly*, it highlights the problem of accountability and corruption in the NGO sector. There are hundreds of NGOs like SDS in Bangladesh which can work due to loopholes in the laws of the country and a lack of sincerity among donors and government. *Secondly*, some people, like the ED of SDS, have taken NGOs as their business and exploit their workers as many businessmen do. *Thirdly*, if field workers are exploited like Mr. Rahim, how could they help their clients to become free from exploitation?

Field workers should be respected more. Goetz (1995, 1996) identified the women field workers of Bangladesh as social pioneers. She also called them 'local heroes' (Goetz, 1996). With the changes women field workers make and the struggles they have, it seems the right term for the thousands of women field workers of NGOs in Bangladesh. Kirlels (1990) sees women field workers as professionals since they have to fulfil their organisations' expectations. So, what should I say after discussing the personal lives of the field workers in the last Chapter and their professional lives in this Chapter? Field workers (both men and women) are simultaneously social pioneers and professionals. The degree of their pioneering in their society depends on individual skill, education, motivation and the opportunities offered by their NGOs. Obviously, the scope and area of work vary from small 'partner' NGOs of SCF to large NGOs like PROSHIKA.

Field workers of NGOs in Bangladesh are social pioneers because they are bringing changes to the lives of their clients and breaking age-old social conventions by working in rural areas, riding bicycles and motorcycles and working in those remote areas where government bureaucrats or staff would never go. They are social pioneers and professionals even with all their personal and professional problems in the field. With the explosion in numbers and outreach of NGOs, field work in NGOs is now a profession and increasingly educated men and women are doing it. Maybe they are doing field work as a profession after failing to get government jobs, but it is still a profession like many others. So, I repeat, how good a social pioneer they are or how professional they are depend on many factors. A major factor is how field workers are supervised or controlled by their superiors and dealt with by their clients. In the next Chapter I shall discuss the interaction between the field workers and their clients and superiors.

Chapter 7

Interactions between Field Workers and their Clients and Superiors

In this Chapter I shall deal with the interaction between field workers and their clients and between field workers and their immediate superiors. This discussion is important for three main reasons. *Firstly*, it is important to know how field workers interact with their clients and their immediate superiors. *Secondly*, from a 'development' point of view, it is important to see how policies and decisions are filtered down from the top to the field workers and how field workers pass them on to their clients. *Thirdly*, it is very important for the NGO management to know how or whether the problems or opinions of the clients are taken into consideration in the short or long-term planning of the NGOs.

On the one hand, I wanted to see what clients think about their field workers and what field workers think about their clients. On the other, I wanted to explore what field workers think about their immediate superiors and what their immediate superiors think about their subordinates i.e. the field workers. With what type of client do field workers prefer to work? What type of field workers do clients like? In the case of field worker/superior interaction I sought to explore the preferences of the field workers and superiors with regard to each other. I wanted also to explore the perceptions of the clients of the services of NGOs. I shall discuss what clients want that field workers cannot give, as well as the achievements and failures of the field workers. These are important aspects of the interactions between the field workers and their clients, because a major role of the field workers is service delivery.

The Client/Field Worker Interface

Throughout my field work I tried to see and know how clients and field workers interact. In most cases, field workers try to order their clients to get their work done. This problem has been exacerbated by the recent emphasis on microcredit. In many cases, to recover the loans, field workers abuse their clients.

From my observation and discussion with the clients, I can grade the NGOs according to their field worker/client relationship. I would put PROSHIKA at the top, and suggest that the main reason for this good relationship is that until recently PROSHIKA was a different type of NGO, more interested in the motivation and organisation of the landless. I have discussed the change in PROSHIKA's goals and working methods. Although PROSHIKA has recently joined the fashion for microcredit, its training and some of its work still leave room for greater interaction between field workers and clients. By 'some of its work' I refer to the distinctive nature of PROSHIKA's work in motivating its clients on local and national political issues. (I shall discuss the pros and cons of this political role of NGOs in the next Chapter). PROSHIKA still runs nonformal and adult literacy schools. PROSHIKA seeks to increase motivation through an emphasis on national pride. I found PROSHIKA field workers were required to observe national days; this is counted in their staff evaluation. In the field I found that all field workers and trainee clients, led by the AC, place a wreath at the Language Martyrs[1] memorial at midnight on 20th February. Very few NGOs in Bangladesh give such importance to observing national days. In the 1980s PROSHIKA was engaged in motivating its clients to demand state land and water bodies for the landless. There are other reasons too. Many of today's senior managers of PROSHIKA were trade union leaders or activists of the left or centrist political parties before joining PROSHIKA. As time has passed, few have changed completely, as this would be difficult. In my experience, leftist political or cultural workers are good motivators. Many senior and mid-level managers of PROSHIKA lamented to me that as the NGO moves away from this motivation, the amicable relationship between the field workers and clients is gradually disappearing. As Hulme and Mosley (1996) argue, NGOs are no longer vehicles for social mobilisation to confront existing socio-political structures.

After PROSHIKA, the clients of the SCF (UK) 'partners' seem to have good relations with their field workers. There are several reasons for this. *Firstly*, I have already mentioned the background of these women field workers. So, clients deal with their field workers with respect and keep in

---

[1] There is at least one in every thana town. The national days of Bangladesh are celebrated on 26th March and 16th December commemorating the start of the liberation war and victory respectively. The functions of the Language Martyrs day and the national days are usually organized by the state at all administrative levels and by other socio-cultural organisations.

mind the influence of their families. *Secondly*, the kinship relation between clients and field workers is also a decisive factor. Clients told me that they like their field workers because they were from their own or neighbouring villages. One client told me 'she is our girl, how good or bad she is is of secondary importance'. *Thirdly*, I have mentioned that SCF (UK) started work in Naria as a relief agency and many people in that area gratefully acknowledge the benefits they received then. There is a dark side too. Here again, microcredit programmes in many cases lead to bitterness between field workers and clients.

After PROSHIKA and SCF 'partners', I would put the relationship between clients and field workers in MCC in third place. Again, there are several reasons. *Firstly*, the missionary nature of the NGO and its activities is a major reason. All the clients described MCC field workers as well-behaved and punctual. As a policy, MCC places high importance on the punctuality and behaviour of its field workers. *Secondly*, a major advantage of the MCC field workers over most other NGOs is the low priority of microcredit and a different way of working with it. So, they are removed from the bitterness of working on microcredit. At the same time, the low priority of microcredit is also a disadvantage. Many MCC field workers told me that clients want microcredit and in some cases leave MCC for credit-giving NGOs. A major complaint I heard from MCC clients is: 'can we eat latrines or health lessons? We need money to survive'. MCC field workers cannot give as much microcredit as their clients demand, so they have a different problem with microcredit.

Among my four study NGOs I found the worst relationship between clients and field workers in RDRS. It may organise picnic and sports tournaments for its staff but it clearly has problems in the microcredit programme. Another issue is that RDRS has failed to re-orient itself successfully from a relief agency to a 'development' agency. Clients do not like to accept that they must now repay RDRS for services. Also, a paucity of skilled field workers in many areas has resulted in poor service delivery by this NGO.

It is essential to have an amicable relationship between clients and field workers. Where it is present, it makes the work of the NGO easier and service-delivery improves. Apart from the issues discussed above, I found that clients want field workers to be polite, co-operative and sympathetic to their problems. All women clients prefer to work with women field workers (despite class differences) and the reasons are quite obvious. In many cases clients want something from their field workers which may be very important but not obvious to outsiders. For example, women clients generally want field workers to be modestly dressed, and, particularly in the

MCC working areas, to maintain purdah. I found most women and men clients to be satisfied with the style of dress and the deportment of the women field workers. This is important, as one RDRS woman client told me that some women in her village complained to her about a woman field worker who 'laughs loudly'. Similarly, all women and men clients want their men field workers to pay respect to the elders of their community and the purdah of their women.

I observed all the married and unmarried women field workers of MCC and SCF 'partners' to be wearing the saree,[2] while some unmarried women field workers of PROSHIKA and RDRS wear the salwar-kameez. Interestingly, all women and men clients told me that they want to see their women field workers in a saree. The saree is the most respectable dress for Bangladeshi women. Women field workers told me that it is more convenient to ride a bicycle or motorcycle and to move around in the salwar-kameez than a saree. Those field workers who wear salwar-kameez told me that women clients regularly ask them to wear a saree and say that they look better in that dress, and that the saree is the appropriate dress for women. Some women field workers who do wear the salwar-kameez told me that they wear a saree for functions and meetings with their clients. Some told me that, although they wear the salwar-kameez, they keep their sarees in their bags to put on for meetings and functions.

Clients not only respect their field workers, I found them sympathetic to them too. Most clients told me that they do not like their field workers (particularly women) to work in inclement weather. Most field workers gratefully acknowledged this but told me that they have no alternative because they are bound by the policies of their NGOs.

Motivation of Field Workers

There are certain things which motivate the field workers to work better. These include the policies of the NGOs, how they are implemented and team work among the field workers and their superiors. All these affect the field worker/superior relationship and the motivation of field workers to work. I have set them out in Table 7.1. The table highlights how NGOs should deal with their field workers and keep them motivated.

---

[2] The saree is a single long piece of cloth wrapped skilfully around the body. The salwar kameez is a long-sleeved tunic worn with loose trousers. The saree is more formal and elegant, but more restrictive. In some areas of India, girls wear the salwar kameez.

Table 7.1 Motivation of Field Workers

	MCC	PROSHIKA	RDRS	SCF 'partner' NGOs
Code of Practice (positive)				
Promotion policy and its implementation	Poorly accepted	Moderately accepted	Poorly accepted	Not applicable
Transfer policy and its implementation	Poorly accepted	Moderately accepted	Poorly accepted	Not applicable
Code of Practice (negative)				
Grounds for job loss	Poorly accepted	Poorly accepted	Poorly accepted	Not applicable
Penalties	Poorly accepted	Poorly accepted	Poorly accepted	Not applicable
'Show cause notices'[3]	Poorly accepted	Poorly accepted	Poorly accepted	Not applicable
Criticism by superiors	Depends on case	Depends on case	Depends on case	Depends on case
Head office visits	Accepted	Accepted	Accepted	Accepted
Team spirit				
Social events	Absent	Highly accepted	Accepted	Highly accepted
National days	Not granted	Highly accepted	Accepted	Not granted
Personal				
Help in emergencies	Accepted	Accepted	Moderately accepted	Accepted
Leave	Poorly accepted	Poorly accepted	Poorly accepted	Poorly accepted

Source: Field Survey.

If we see how NGOs can satisfy needs at different levels of Maslow's hierarchy (from Morgan, 1997, p. 37) it is clear that NGOs do not feel it important to meet the needs due to the absence of job security and prevalence of patronage in these organisations (see Table 7.2).

---

[3] Show cause' is disciplinary action taken by NGOs asking their staff to defend themselves in writing against activities like insubordination, corruption, irregularity, misconduct etc.

Maslow set out a hierarchy of needs (Table 7.2), as follows. First, physiological needs are the very basic needs such as air, water, food, sleep, sex, etc. Then, safety or security needs have to do with establishing stability and consistency in a chaotic world. These needs are mostly psychological in nature: we need to feel safe. Love and belongingness or social needs are next on the ladder. Humans have a desire to belong to groups: clubs, work groups, religious groups, family, gangs, etc. We need to feel loved (non-sexual) by others, to be accepted by others. Then come two types of esteem needs. First is self-esteem or ego which results from competence or mastery of a task. Second, there's the attention and recognition that comes from others. This is similar to the belongingness level. The need for self-actualization is the desire to become more and more what one is, to become everything that one is capable of becoming. Maslow based a whole theory of humanist psychology on this hierarchy or ladder of needs; here, it is used only to illustrate some aspects of NGO provisions for their field workers.

Table 7.2   How NGOs Satisfy Needs at Different Levels of Maslow's Hierarchy

Type of Need		Satisfaction
Self-actualising	Encouragement of complete employee commitment	?
	Job as a major expressive dimension of employee's life	No
Ego	Creation of jobs with scope for achievement, autonomy, responsibility and personal control	No
	Work enhancing personal identity	Limited
	Feedback and recognition for good performance (e.g., promotions, 'employee of the month award')	No
Social	Work organisation that permits interaction with colleagues	Yes
	Social and sports facilities	Yes (except SCF 'partners')
	Office parties and outings	Yes (except SCF 'partners')
Security	Pension and health care plans	No
	Job tenure	No
	Emphasis on career paths within the organisation	No (except PROSHIKA)
Physiological	Salaries and wages	Yes
	Safe and pleasant working conditions	No

Source: Field work.

Clients' Perceptions of The Services of their NGOs

A major complaint that I heard from most field workers is that clients perceive services from their NGOs as given free or as 'relief'. In MCC's Soybean Project area, all soybean producers were telling me that MCC should buy their soybean since they were facing losses from its low market price. To popularise soybean at the initial stages of the project, MCC distributed free seeds and bought soybean from the farmers at a very high price. Now MCC distributes soybean seeds at a subsidised price and organises cooking demonstrations to popularise soybean as an important crop.

RDRS clients and field workers told me that RDRS distributes some magazines and education materials at nominal prices but clients demand them free. Many RDRS clients complained that RDRS is making money for things which were free earlier (when RDRS was an international NGO). Some RDRS literacy magazines are priced at Tk 3 each and field workers are ordered to sell at least one copy to each group. The record books for groups are printed by RDRS and groups have to pay Tk 100 for each. All RDRS field workers told me that they face immense problems in getting money for these from their clients.

NGOs like MCC, RDRS or SCF (UK) start relief work after natural hazards and work as relief agencies. Clients still perceive the services of their NGOs as free. A culture of relief if is really a problem in promoting 'development' or self-reliance. Field workers blame the NGOs for this poor planning.

A major service provided by the NGOs to their clients is training. The main complaints from the clients on training were poor timing and some irrelevant training. Due to the loss of wages for days or weeks, clients are unwilling to go on training programmes. One important problem is the poor timing. All clients (both men and women) told me that there are times during the year when they cannot afford to be away from their work, for example, during the sowing and harvesting seasons. All clients told me that this problem could be solved by asking them when are the best times. Some training is very popular with the clients because they find it useful, for example home gardens, record-keeping, poultry rearing etc. Clients told me that some training is not useful and they find it a waste of time. When I asked some clients what they had learned from training which they took six months or a year ago, many could not tell me. Except for SCF 'partner' NGOs, all the study NGOs have regular training programmes for their clients. (SCF (UK) arranges visits to other NGOs but only for its 'partner' NGO field workers and directors, not for clients). MCC and PROSHIKA clients get transport and food allowances and accommodation for training programmes, which seemed to compensate adequately for their absence from work if the timing was convenient. RDRS clients, on the contrary, have to pay Tk 25 for three days training and Tk 50 for training lasting more than three days but their clients dislike this system and gave me the examples of other NGOs. All RDRS clients told me that they cannot afford these training fees and leaving paid work or work in their farm or home, when they find it difficult to repay their loans anyway. Many RDRS clients told me that they are very often forced by their field workers to go for training. Above all, they complained to me about the poor quality of food in the training centres. Most RDRS field workers interviewed told me that if their clients do not go for training in large numbers, their district or head office will demand to know why. So, they sometimes force their clients to go for training.

A major complaint that clients made to me about their field workers is the lack of punctuality of some. Sometimes field workers come late to group meetings, which annoys many clients who have left paid work or domestic responsibilities to attend. I found field workers sometimes to be late for unavoidable reasons, but also in some cases to be really uncommitted or not serious about the value of their clients' time. Sometimes mid-level managers, when they announce they will meet the clients may also be late too. When I was working in Companyganz Thana of the MCC working area, all the women clients complained to me that a woman anti-diarrhoea campaign worker accompanied by the hex (woman) senior failed to come to their village on the pretext of rain while all the clients waited for her for the whole day. In these cases, field workers bear the brunt of these grave mistakes of their superiors and have to face the criticism of their aggrieved clients.

I would like to close this section with a discussion on the opinions of the clients about the services they need from their NGOs, in order of priority.

1. *Unity*: All clients told me that they want advice and supervision from their field workers on how to maintain unity in their groups. This seems a good suggestion, because NGOs should work to make their client groups self-sustaining so unity among the clients is essential. Most clients told me that members do not all join the groups with the same objective so it becomes difficult to maintain group unity when some shrewd people join the groups to fulfil their petty interests.

2. *Credit*: This may seem unexpected as I have been critical of credit but when I was talking with the clients all of them told me that to make a

living everybody needs money, or more precisely a regular income. Clients say that credit should be given due importance as a service for them.

3. *Education*: All clients told me that they want nonformal and adult literacy services from their NGOs. Some told me that after completion of these education programmes there should be separate programmes to retain this education otherwise they would forget their reading and writing skills which would be a waste of time and resources.

4. *Skill training*: Most clients (both men and women) gave high importance to skill training. They told me that to become self-employed or to use credit efficiently they need to be skilled in their trades. So, clients need skill training for income generating activities.

5. *Gender awareness programmes*: All the women clients, and even some men, told me that the gender programme of the NGOs could not be successful without legal awareness programmes on the rights of women, the necessity of girls' education and awareness creation against child-marriage and dowry. These, they say, are not only necessary for women but for men too.

6. *Health-education*: All clients told me that they need health education programmes like cleanliness, basic health awareness, popularising the necessity of using latrines and supplying them at subsidised prices, sinking tube-wells and making clients aware of the necessity of using safe drinking water etc.

The above suggestions require proper attention from the policy makers of NGOs in Bangladesh. I shall discuss the field workers' opinions on services in the next Chapter.

Field Workers' Opinions of their Clients

So, with what type of clients do field workers want to work? They tend overall to prefer women, not-so-poor, educated and obedient clients. Women are clearly more obedient than men, as I shall discuss below. My findings are similar to those of Rahman (1999) on the Grameen Bank. Table 7.3 shows the field workers' preferences for clients. The gender variation in selecting clients by the field workers does not seem unusual. Clearly, all women field workers would like to work with women clients. Interestingly most men field workers (63 per cent) wanted to work with women clients while 26 per cent wanted to work with both men and women.

To elaborate on why women clients are preferred to men, the field workers' reasons[4] are given below.

1. Same gender (for women field workers only).
2. Women are always in the house, so always available for a meeting.
3. Men create problems in repaying loans, women are better. Men are more ingenious and difficult to control.
4. Women are committed to repaying loans, utilise loans properly, and usually do not waste money.
5. Women try to be self-reliant.
6. Women attend meetings regularly.
7. It is easy to work with women.
8. Women are more united than men and groups do not break up.
9. Women are good savers.
10. Women are obedient.
11. Those women who know that their husbands will misuse the money do not take loans.
12. Women do not have the courage to flee the village.

Table 7.3   Field Workers' Preferences for Clients

Name of NGO	Both	%	Women	%	Men	%	Total	%
MCC	19	65	13	19	10	90	42	39
PROSHIKA	8	28	8	12	0		16	15
RDRS	2	7	33	48	1	10	36	33
SCF 'partner' NGOs	0		15	22	0		15	14
Total 100%*	29		69		11		109	

*Percentages are rounded so they may not add up to 100 per cent.

Source: Field Survey

The list above gives rise to several questions. A major question that strikes me is that not a single field worker (man or woman) told me that they prefer to work with women because they need 'development' or empowerment, perhaps reflecting a mentality which puts their own convenience first. This also raises questions on the quality of training of the field workers.

---

[4] All these were mentioned by several field workers.

My findings also confirms the findings of Goetz (1995, 1996, 1997) and Goetz and Gupta (1994) which questions the quality of the Gender and Development programmes of NGOs and GOs where women are used as a means to provide credit to a family. When I was talking to Rina Sen Gupta she told me, 'if NGOs want to give credit to men they should give it directly to men and why are they using women as a medium?' (Dhaka, April, 1998). The way NGOs give emphasis to credit programmes targeting women seems in most cases to be a channel to give money to men. Kabeer (1998) rather differs with Gupta and is a supporter of the present system.

The field workers' and mid-level managers' preference for women clients can be elaborated by an example. In Kurigram Sadar Thana, RDRS had around 4000 members of whom 2600 were women and 1400 men. The Thana Manager of Kurigram Sadar told me that he had directed his field workers not to take any more men clients and to enrol only women. He even told me that if men clients wanted to leave RDRS to join other NGOs, field workers should allow them to leave but try to keep the women clients (December, 1997).

Field workers' preferences for not-so-poor, educated clients could be linked to the microcredit programmes of the NGOs. Even in MCC, which does not work so much in microcredit, field workers told me that to work in agriculture clients need some land. Many MCC field workers expressed to me their concern for the landless people who could not be reached by their NGO. I shall discuss the problems of microcredit in the next Chapter. Here I would say that it is convenient for the field workers to work in microcredit with the less poor, who are good repayers. The preference of the field workers for educated clients once again represents the problems of the NGOs mentioned above. Educated clients understand arithmetic well, usually do not create trouble over accounts, understand everything easily and can help in keeping the records of their groups. But if the NGO priority is to target the disadvantaged and mobilise the poor, the preference of the field workers for educated and less poor clients raises a big question.

The main problems of working with men and women clients as reported by the field workers[5] are listed below:

Women

1. Husbands create trouble (do not allow them to join groups, come to the meetings, force them to ask for loans etc.).
2. Problem of illiteracy.
3. Lack of interest in going for training.

---

[5] Again, all were mentioned by several field workers.

4. Difficulty in being contacted by opposite gender (for men field workers only).
5. Early marriage.
6. Afraid of NGO being Christian.
7. Problems in coming to meetings due to child-care and other domestic responsibilities.
8. Hesitate to discuss family planning (for men field workers only).

Men

1. Do not repay loans regularly.
2. Migration.
3. Dishonest people create trouble.
4. Misuse the loan.
5. Illiteracy and lack of education.
6. Afraid of NGO being Christian.
7. Want quick credit.

The above list highlights some key issues. For example, there are some common problems for both men and women like illiteracy, or fear of religious conversion. The above list underlines the necessity of providing education programmes (both nonformal and adult) by the NGOs. There are some typical problems in working with men, such as dishonesty, seasonal or permanent migration etc. Interestingly, the problems in working with women clients are mainly caused by their husbands or society e.g. restrictions created by husbands, domestic responsibilities and early marriage. These problems could be changed through more awareness-creation among both men and women. Unfortunately, most field workers told me with frustration that they have very little time or support from their NGOs for awareness creation.

What Clients Want Which Field Workers Cannot Give

There are certain materials and services for which clients ask their field workers, who cannot provide them. These deserve a mention because all field workers told me that they feel helpless when they find some services essential for their clients but cannot give them, due to the limitations of their NGOs. The services are:

1. *Contraceptives*: All women field workers told me that they regularly get requests from their women clients for contraceptives. There are

several reasons for this. *Firstly*, most women clients prefer contraceptives to sterilization but due to purdah and other cultural constraints women cannot directly buy contraceptives as these are sold by men. *Secondly*, women cannot go to the markets by themselves to buy contraceptives and the state family planning workers rarely visit their homes (as discussed in the last Chapter). Women cannot go to the shops or send men other than their husbands to buy contraceptives for them due to shame. So, the only option left is to depend on their husbands. The problems with husbands, mentioned by women field workers, are first that some husbands are not as interested in family planning as their wives (more prevalent when they have not yet fathered a son); second, some husbands are not serious about family planning so they forget to buy contraceptives regularly for their wives; and third, there is shame. Some women field workers told me that on many occasions their women clients offer them money to buy contraceptives for them and sometimes they (the women field workers) do so. But in most cases women field workers find it very difficult to convince their clients that they cannot provide free contraceptives or buy them from the market due to their time constraints and the limitations of their NGOs. One SCF 'partner' NGO field worker told me that one client had been asking her to buy contraceptives for three months but she could not manage it for her. Recently, the field worker had discovered that her client had become pregnant (November, 1997). Most women field workers pointed out to me that difficulties in access to contraceptives for women are a major barrier to family planning programmes in Bangladesh. When I asked women field workers whether they think that their NGOs should start distributing contraceptives they told me that with the workload they had it would not be possible. Instead they suggested starting awareness creation programmes among men for family planning.

2. *Nonformal education*: Among the study NGOs only PROSHIKA has a nonformal education programme for children. Field workers of the three other NGOs told me that clients regularly ask them to start nonformal schools for their children. Many field workers lamented to me that clients have been made conscious about education but could not get the service for lack of schools. When I asked clients why they did not send their children to the government primary schools they gave me several reasons. *Firstly*, clients and field workers told me that the distance to the schools, poor infrastructure, lack of seating facilities, frequent absences of teachers and unattractive education methods discourage students from enrolling or make them drop out. Dreze and

Sen (1995) found a similar situation in India. *Secondly*, although in theory primary education is 'free', it is not so in practice. Clients and field workers say that teachers charge for books, examinations and functions etc. *Thirdly*, when children grow up and want to go to school they are ashamed to sit in classes with younger children, so nonformal schools are the only choice for their education. I discussed the advantages of the PROSHIKA nonformal schools. I found PROSHIKA nonformal schools to be free and teaching methods both attractive and more life-oriented than in state schools. I found the nonformal schools much better than the state schools in all respects. All PROSHIKA nonformal schools have inspection books and when a mid-level or senior manager or visitor like me visits the schools they can record their opinion on cleanliness and quality of teaching.

3. *Medicines; salt for oral rehydration therapy (ORT)*: Clients ask for these materials mainly for two reasons. *Firstly*, due to awareness created by the NGOs, many clients feel that it is necessary to take in saline[6] when they are affected by diarrhoea. They also think that due to vitamin deficiency their children need to take vitamin capsules. *Secondly*, some NGOs distribute free ORT salts and vitamin capsules when they get them from donors. So, many clients think that their NGO should provide them with free salts and capsules too. Sometimes this creates misunderstandings between clients and field workers. State health workers are supposed to distribute salts free but usually do not. Many women field workers told me that they show their clients the simple procedure of making rehydration drinks with raw sugar and salt at home, but clients prefer to get ready-made packets to make up. All field workers were against the distribution of salts because they think it is very easy to make ORT drinks at home. The field workers told me that provision of these materials would create dependency between the NGOs and their clients.

4. *Tailoring training*: Only RDRS have a system of providing training to their clients for tailoring. I found a well-run tailoring training programme for women clients of RDRS. Initially RDRS provides the sewing machines, cloth and other materials and trains its women clients. After three months training clients start to earn money, from which they can repay the costs of machines. Many field workers told me that there are two main reasons for this high demand for tailoring skills. *Firstly*, tailoring is a very convenient trade for women. Women can work on tailoring at home, which does not affect their purdah and

---

[6] Salts for ORT.

they can do it with less effect on their domestic responsibilities. Above all, women usually like to order their clothes from women tailors. *Secondly*, tailoring training helps women to get jobs in the mushrooming garment factories in Dhaka and Chittagong, which mainly employ women. These reasons strongly call for the adoption of tailoring training programmes by the NGOs in Bangladesh.

5. *Credit*: Requests for this service only come from the MCC clients. MCC field workers told me that their clients ask for credit, citing the examples of NGOs such as BRAC, ASA and organisations like the Grameen Bank. Many clients become frustrated when they can not get credit as from other NGOs. Sometimes these disenchanted clients leave MCC membership. MCC operates its credit programmes on a small scale from the savings of the clients. MCC clients want MCC to run its credit programme like other NGOs. All MCC field workers told me that they try to convince their clients of the dangers, showing the examples of BRAC or Grameen Bank clients, some of whom sold their cattle, ornaments or assets to repay their loans.

6. *Tube-wells*: This is a costlier service than those discussed above. Most field workers told me that during health awareness programmes they motivate their clients to use tube-well water. Many clients ask their field workers to sink tube-wells in their villages. RDRS subsidises the sinking of tube-wells and SCF (UK) sank many tube-wells before the handover. Interestingly, no field workers (including those of RDRS) liked the idea of providing tube-wells because they thought it could be done by the clients themselves. There is another major problem too. Recently, there has been an outbreak of arsenic contamination in the tube-well waters in many parts of Bangladesh. This has made the work of field workers, who earlier encouraged their clients to use tube-well water, difficult. They now have to advise their clients not to use it. Most field workers told me sadly that they feel themselves to be a laughing stock for changing their advice.

The Interaction between Field Workers and their Superiors

I have discussed the choice of field workers by the senior and mid-level managers of NGOs. I repeat, senior and mid-level managers prefer obedient, sincere and intelligent field workers. If I were in their place I might have said the same thing. For example, one SCF 'partner' NGO director told me that she does not like those field workers who fail to book their leave at least two-three days in advance because it hampers her work.

Briefly, the relationships between field workers and their immediate superiors vary from NGO to NGO and person to person. Overall, I found the relationship between the directors and field workers of SCF 'partners' to be the best, in the sense that trust is greater and they are more sympathetic to each other. The directors and most field workers were colleagues who formed their own NGO and they are from the same area. The relatively flat structure of these NGOs could be another reason for the good relationship between the field workers and their superiors.

After the SCF 'partners' I would rank the PROSHIKA field staff/manager relationship, which might seem unexpected because of the huge size of this NGO. The major reasons behind the good relationship between PROSHIKA field workers and their immediate superiors are better management at the top and mid-level, and the clear system of promotion. But I have heard some examples of bitter relationships between PROSHIKA field workers and mid-level managers too. After SCF 'partners' and PROSHIKA I would put the staff/superior relations in MCC and RDRS at the same level because I have heard and seen many examples of bitter relationships in those NGOs. I shall begin with the good relationships between field workers and their superiors where it is best and go on to the poorer relationships, in order to show how relationships between field workers and their superiors become bad and how they could be improved. When I was talking to the mid-level and senior managers of PROSHIKA they gave me some suggestions for keeping good relations between field workers and their superiors. These include:

1. If the superiors find any problem with the field workers they should not report it to a higher authority or serve 'show cause'[7] notices to the field workers concerned. I heard many complaints of 'show cause' or disciplinary action by MCC and RDRS mid-level managers against their field workers. PROSHIKA mid-level and senior managers take official action only as the last resort against a field worker. Instead they sit with the field worker and try to solve the problem through discussion. One Zonal Coordinator of PROSHIKA told me that he did not, even verbally report any complaint against any field workers to his superiors when he was Area Coordinator (AC). He always preferred mutual discussion and he encouraged all his ACs to do that.

2. Another way of correcting the errors of field workers is to discuss these with them in formal and informal meetings without mentioning any names. Instead they say, 'We should not do that'.

---

[7] 'Show cause' is disciplinary action taken by NGOs asking their staff to defend themselves in writing against activities like insubordination, corruption, irregularity, misconduct etc.

3. Where there is a good relationship between field workers and their superiors, the key element is team-spirit. Some mid-level and senior managers told me they look after the personal problems of the field workers like accommodation, health problems or financial problems. Woolcock (1998) reported that he found a mid-level manager in Grameen Bank assuming the roles of marriage counsellor, conflict negotiator, training officer, civic leader and bank manager. I found some mid-level managers of PROSHIKA to be like this.

4. All field workers and their superiors with whom I talked told me that their relations are best where superiors are sympathetic. I found some mid-level and senior managers to be really sympathetic to their field workers, and their generosity seems to pay-off (compare Palmer and Hoe, 1997). However, I have also seen that most senior and mid-level managers believe in strict supervision of their field workers. After talking to the field workers and managers my advice would be to be more sympathetic, as the following discussion will illustrate.

Sufia Begum, Development Worker, PROSHIKA, Sakhipur: Women field workers in PROSHIKA get maternity leave with all benefits from six weeks before delivery until six weeks after. Sufia did not take her leave before delivery, although the Area Coordinator (AC) and Zonal Coordinator (ZC) both asked her to do so. One day, she had pains on her way to the field area and her clients rushed her to the PROSHIKA office. The AC managed to get an ambulance for her, which is really difficult in rural Bangladesh, and took her to the Tangail District hospital, where medical facilities are better than Sakhipur Thana Health Complex. Sufia gave birth to a son. The AC bought the medicine for Sufia and she repaid the money in six months. Soon after the delivery, two PROSHIKA field workers went to inform Sufia's husband (who was in Mymenshingh) of the news and took him to the hospital on their motor cycle. All the staff from Sakhipur and Tangail regularly visited Sufia at the hospital. One month after the birth, Sufia's son got an infection and Sufia informed her AC about her son's ailment from her village home at Tangail. Although she was supposed to return to work within six weeks after the delivery, due to her son's illness she returned after three months. After her return the AC and ZC both advised the head office to consider her absence from duty as medical leave, which was approved. Sufia gratefully acknowledges the sympathy of her colleagues and superiors who did not ask her do any heavy work before her and her son's full recovery which took eight months (February, 1998).

When I was in Sakhipur two field workers (one Thana Coordinator and one Development Worker) were transferred to other Areas and two new field workers took their place. The farewell and arrival of the field workers was marked by a function and a dinner. Officially, PROSHIKA usually discourages these functions. But I saw the enthusiasm of the field workers and the AC and ZC who all paid for the dinner and gifts for the four field workers. I was informed that this kind of function is a long tradition of PROSHIKA, where all field workers and their superiors and their spouses (in some cases children) take part. All these activities create a friendly relationship with and between the field workers. I have not seen or heard of any social activities like this in the three other study NGOs.

I observed and heard about many cases of bad relations between field workers and their superiors.

Some cases of bad relations are created primarily by field workers themselves, some by their superiors. The major causes appear to be: some superiors are too eager to discipline or control their field workers, which creates misunderstanding; some superiors receive certain benefits but withhold them from their field workers and this creates frustration; above all, as I have mentioned earlier, some superiors are not sympathetic to their field workers' problems and grievances, as I shall now illustrate.

Abdul Quyyum, field worker, Kurigram Sadar, RDRS: Mr. Quyyum and four other men colleagues were living in a mess, which was cheaper than living individually. Also, they were enjoying the association of their colleagues by living together. They decided not to hold any group meetings after dark due to the risks of carrying money unless a senior colleague took them on a motorcycle. Mr. Quyyum and his mess-mates decided to unite to fight for this. In one staff meeting Mr. Quyyum and his colleagues raised the issue and their superiors immediately turned down the proposal saying it would hamper overall repayment and men clients prefer to meet after dark. The men field workers said, due to lack of security, the money should be carried by motorcycle.

Mr. Quyyum thinks their superiors hatched a conspiracy against them. First of all, the superiors broke up their mess by getting a local youth to write a complaint against one of his mess-mates about an illicit relationship with a local girl. Mr. Quyyum and his colleagues told the Thana Manager (TM) that the complaint was baseless and written out of enmity. But the TM did not accept their defence. He said that if they break up the mess he would not send the complaint to the higher authorities, which would result in disciplinary action against them. Mr. Quyyum thinks that he and his colleagues were not only harassed by their superiors; they were affected financially too. Now he spends Tk 1000 per month for food and accommodation which cost Tk 500-550 when he was living in the mess.

Mr. Quyyum told me that since the dispute with his superiors, some do reluctantly go with them to the meetings after dark. On another occasion Mr. Quyyum found (in the middle of December) that he had one day of casual leave left. As per the rules of RDRS if he did not take the leave, it would be lost. So, he asked for one day's leave but both his superiors refused, saying that collection of money would be affected if he took leave then. He earnestly pleaded with his superiors because his wife was sick. After prolonged requests his superiors allowed him to go home. At his village home Mr. Quyyum found his wife very ill and stayed for two days instead of one. When he returned to Kurigram his superiors served him with a 'show cause' notice for unauthorised absence for one day. In his reply, Mr. Quyyum apologised for his wrongdoing and replied that his wife was very sick so he could not return to his office in time. In these cases mid-level managers could be more lenient. When I was talking to senior thana staff in RDRS about this kind of event they preferred stringent management because they think leniency makes field workers arrogant and deceitful (December, 1997).

A major problem in the relationship between the field workers and their superiors in RDRS is the tendency among many field workers to get new postings without reference to their immediate superior. I have already discussed the transfer problem of field workers in the last Chapter. All senior thana staff told me that they feel threatened when they find their field workers use 'connections' at the Dhaka or district level offices for their transfers. This problem could easily be solved through the enforcement of strict management at the top level of the NGO.

The mistrust and poor relationship between field workers and their superiors not only affects the smooth functioning of NGOs, it also affects the implementation of their policies. RDRS clearly shows the problem. In Kurigram Sadar some complained to me that when they demanded refreshments in the staff meetings, their superiors all said that there was no provision for that. Some men told me that they went to the accounts section of RDRS at Kurigram District town and found out there was a budget for refreshments in staff meetings. These men field workers decided to submit formal representation further up, but no women and not all men agreed so the matter was abandoned. The aggrieved men told me that they could not speak out against the corruption of their superiors due to their women and less qualified men colleagues who did not want to be disobedient. This example not only shows how the relationship between field workers and their superiors can deteriorate but also illustrates the corruption of mid-level managers of NGOs in Bangladesh.

Making Comparisons

I have already mentioned the lack of sympathy among some women superiors to their women field workers. Many women field workers, particularly of MCC, complained to me that their women superiors are not sympathetic. The conventional wisdom of recruiting women managers for the welfare of women field workers can still be easily challenged. Some women field workers told me that men superiors are in many cases better than women. Many women field workers of MCC, PROSHIKA and RDRS told me that if their women superiors find them not to be in the field or discover any irregularity, they never hesitate to send a memo asking for explanations, whereas in some cases men superiors ask for explanations verbally rather than officially.

So, what happens when field workers report problems to their superiors? Of course, when superiors are sympathetic and the relationships between field workers and their superiors are good, a solution to the problem is to be expected. When the superiors are not sympathetic then problems start. These include:

1. Some field workers told me that when they speak out about their problems, they are identified as 'problem staff' and are disliked by their superiors. In many cases these 'problem staff' are harassed and deprived of promotion or suitable postings by their superiors.
2. Most field workers told me that their unsympathetic superiors tell them 'other people are working with it, if you cannot do with it then leave the job'. Field workers told me: 'what should we and our families eat if we leave the job?'.
3. Some field workers told me that some of their superiors do not want to listen to their professional problems or help to solve them. Instead, they (the superiors) abuse the field workers about problems even though they were beyond the control of the field worker. One RDRS field worker told me that one member of his group sold the group's shallow tube-well and the other members of the group filed a case against the defaulting member. When the field worker reported the problem to his superiors they were very angry at him saying: 'Don't you know that we cannot take any legal action, RDRS cannot be a party in a criminal case like this. Why do you create trouble?' The field worker told me that the group members asked local leaders to exert pressure on the defaulting

member to pay the price of the sold tube-well but failed. So, going to court was their last resort (Zulfiqar Ali, UO, Kurigram Sadar, December, 1997). Other field workers told me that they complained to their superiors that they could not find time to maintain the accounts of their groups. Their superiors told them to take educated people as clients who would be able to do the work. Field workers told me that it is difficult to find educated clients and sometimes they are clever and cheat. Above all, educated clients should not be the targets of NGOs who work for the poor. I have heard many similar complaints.

4. There are some worse cases too. Some women field workers of MCC told me that an overseas woman gender advisor came to visit them and they told her their problems. The women field workers told the advisor about their problems of child-care and accommodation arising out of the MCC rule of living within their working areas and the lack of sympathy from some of their superiors. By some means, their superiors found out about this and threatened them, forbidding them to complain to the advisor. The women field workers told me that the gender advisor asked them for a written complaint, but for fear of losing their jobs and lack of unity among themselves they could not do that. Two women field workers who led the move were sacked on the grounds of 'lack of funds to keep them' in MCC.

In Chapter Three, I discussed the problems that SCF field workers faced during the handover and gave my view that the handover of activities is an example of poor planning for an international NGO like SCF (UK). The relationship between the mid-level managers and field workers plummeted to the lowest level, not only due to the loss of jobs and drastic falls in the salaries of the field workers, but also for other reasons, including:

1. SCF 'partner' NGO field workers told me that they did not believe that 'higher operating costs' led to the handover. They blamed the misuse of money by the SCF (UK) management, giving the examples of buying costly furniture unsuitable for remote rural areas and locating the office in an erosion-prone area which was destroyed by the Padma etc.

2. SCF 'partner' NGO field workers told me that their superiors showed them advertisements in the newspapers for field workers with BA or MA degrees at a Tk 1000 monthly salary, while SCF (UK) was paying its field workers around Tk 4000-5000. Field workers complained to me that they were not allowed to ask their superiors to compare their own salaries with other national and local NGOs. Undoubtedly, they said, SCF (UK) mid-level managers were getting higher salaries than

other national and regional NGOs. All field workers told me that they were very unhappy at this undemocratic attitude of their superiors.

3. A pretext for dismissal given by the mid-level managers of SCF (UK) to its field workers was that the work that was done by ten field workers could be done by six field workers at a much lower cost. All SCF 'partner' NGO field workers told me that there had been very little redundancy at the middle and senior level (I found that the complaint was not totally correct). Field workers told me that they decided to work with the existing staff but with a lower salary.

4. Some mid-level and senior NGO managers exploit the ignorance of field workers. When one SCF 'partner' NGO field worker asked me how many festival bonuses I got, I answered that as a member of staff in a semi-state organisation I got two festival bonuses on two Eids, each equal to my basic salary, and non-Muslim staff get the same. The field workers were surprised, and angry with their superiors, because they had been told that state employees get one festival bonus each year so they should get one.

The field workers gave me many examples of mistreatment by their superiors. Some MCC and RDRS field workers complained to me that if their superiors come to the office or the field late, no action was taken, even if they inconvenienced many field workers. However, if field workers, arrived late or left their office or field early they got a 'show cause' letter. In some of the worst cases, field workers of MCC and RDRS told me that their salaries were cut on the basis of hours they were absent from the office or field. Strikes are a regular phenomenon in Bangladesh.[8] Most MCC field workers complained that when politicians or transport workers call a strike, their superiors just close the office and stay at home, although they could easily get to the office. The superiors who stay at home abuse the field workers if they do the same. All MCC field workers told me that they feel this to be unfair but dare not protest for fear of losing their jobs.

I have already mentioned the general discontent among the MCC field workers with the rule which compels all field workers to live in or near their work area. All MCC field workers were unhappy with this rule. They told me that their supervisors (immediate superiors) also used to have to live in their work areas but, about two years ago, they changed the rule for themselves, which has antagonised the field workers.

Now, I shall summarise the pattern of bad relationship between the field workers and their superiors.

---

[8] Organized by political parties, transport workers, state-owned factory workers etc.

1. In cases where I heard complaints of bad relationships between field workers and superiors, there is always a lack of co-operation from the superiors. Field workers not only want their superiors to be sympathetic to them, they want suggestions and help in their day-to-day work. Where superiors do not co-operate or help, the problems of the field workers automatically increase. Many field workers complained to me that their superiors are unwilling to listen to their problems and instead of solving them they sometimes abuse the field workers for raising problems.

2. Most field workers of MCC, PROSHIKA and RDRS told me that they are close to their clients, and see their problems, but when they report problems to their superiors, the superiors do not listen to their proposals. Some superiors make no effort to grasp the realities in the field (because some of them do not visit the clients regularly) so they sometimes suggest things which field workers find irrelevant. Sometimes superiors undermine the tacit knowledge[9] of the field workers, which makes field workers very unhappy. Let me give an example.

Rahmat Ali and Sikander Ali are cousins and members of a RDRS group. They took a loan from RDRS for aquaculture. Sikander has gastric problems and cannot work regularly. During the repayment period Sikander borrowed money from Rahmat and kept repaying the loan regularly. After repayment of the loan, their group applied for a new loan. But Rahmat asked other group members and his field worker to withhold the sanction of a new loan until Sikander paid back his money. Rahmat and most of the other group members asked their field worker (Jane Alam) to put pressure on Sikander to repay his money. Jane Alam decided not to pass on the documents to his superiors for the loan to be approved until Sikander repaid the money. Mr. Alam complained to me that Sikander went to his superiors and told them that he would repay the money borrowed from Rahmat from the new loan and the Thana Manager approved the loan without discussion with the Assistant Thana Manager and himself. Jane Alam told me that other group members became very angry with him and blamed him for the incident. Mr. Alam told me with frustration that when he told his superiors about his distress they were very angry with him, saying 'We have to maintain a high level of outstanding loans and what is the problem if he repays the loan from the new loan'. Mr. Alam told me that this will create a bad precedent and clients will learn to repay old debts from new loans and

---

[9] The knowledge field workers gain through work.

above all, they will dare to bypass their field workers (Ulipur, December, 1997).

This discussion highlights how bad relationships arise between field workers and their superiors. Some field workers told me they had mixed feelings about their superiors. Many field workers have bad relations with some of their superiors and good relations with others; it varies from person to person. Some mid-level and senior managers are really sympathetic to their field workers but some are not. Many field workers told me that some of their superiors are too much driven by their own self-interest and obsessed with the performance indicators of their NGOs. This kind of superior is obviously unpopular among the field workers. I shall give two examples below.

1. Saidur Rahman, Homestead Development Facilitator, MCC, Maizdee: He thinks his supervisors are not sympathetic to him and his colleagues. Mr. Saidur likes the staff based at Maizdee office who are sympathetic to him, and give useful suggestions and orders. On the other hand, his immediate superiors behave badly to him and his colleagues, making foolish suggestions and creating undue pressure, which he thinks demoralises the field workers. He told me that sometimes clients want to see the staff from the Maizdee office and when, at his request, they visit his clients it works as a booster for the clients (March, 1998).

2. Mohammed Ilias, Education Worker, PROSHIKA, Sakhipur: He joined PROSHIKA in 1981. He told me that he liked the philosophy of PROSHIKA and its senior managers. His job was to organise groups, mobilise clients to demand (in some cases to invade) government land, conscientise the farmers and labourers for rights etc. In doing so, he had 14 criminal cases against him and had to go into hiding. Mr. Quazi Farooq Ahmed, the Executive Director of PROSHIKA helped him to continue his litigation and gave him a prize of Tk 6000. Later, all cases against him were withdrawn due to the sincere efforts of the Executive Director.

In 1987, Mr. Ilias was transferred to Bikrampur from Manikganz in Dhaka District. After the devastating flood of 1988 he had a serious conflict with his superiors, whom he blamed for corruption in the distribution of relief goods. Mr. Ilias lost his job on the charges of insubordination and corruption. But Mr. Ilias thinks he lost his job for three main reasons. *Firstly*, he reported the corruption of his superiors and some of his colleagues to the higher authorities. *Secondly*, his superiors were always sceptical about him because he was the only graduate among the field workers and mid-level managers in the Bikrampur office. *Thirdly*, there were serious internal political divisions in PROSHIKA. He had been a worker for a centre-left political party and joined PROSHIKA with many of his comrades. All field workers from this background were isolated by others in PROSHIKA. As a consequence, he and 48 other field workers simultaneously lost their jobs in 1988. After losing his job, Mr. Ilias got three months salary as a gratuity. Mr. Ilias met the Executive Director (who had recruited him), who gave him a patient hearing and told him to apply for re-recruitment. Mr. Ilias was re-recruited and posted at Mirzapur Thana of the Tangail District. Here again, he fell prey to his opponents but the ED was very sympathetic to him. However, Mr. Ilias is working as a field worker when many of his former colleagues have been promoted as mid-level and senior managers (Sakhipur, February, 1998).

Have Field Workers Achieved What They Want To Do?

I asked all field workers for an evaluation of their own work. Although most policies are formulated at the top level of NGOs, all field workers feel there are certain services which they should give or improve in quality. In this section I shall discuss this and I shall elaborate on the self-assessment given by the field workers.

MCC Bangladesh

MCC field workers mentioned to me achievements such as: a) Some clients got an economic lift through the services provided by MCC; b) Poultry rearing, cattle-raising, homestead gardening and vegetable cultivation and consumption have increased among the clients; c) In general, clients have become more conscious about womens' rights and health (but dowry is still a major problem). MCC field workers told me that generally half of their clients have benefited from the services of their NGO. So, why not the other half? Field workers gave me several reasons. *Firstly*, technical advisers usually help farmers with at least one acre of land, as farmers with less than one acre of land (who are functionally landless) cannot reap benefits due to inadequate physical assets. MCC field workers pointed out to me that the present way their NGO works does not help the functionally landless and cannot reach the landless who are not usually targeted.

One problem identified by the MCC field workers is that many of their clients remain vulnerable to nature, social structure etc. Some field workers told me that they feel devastated when they find their clients uprooted from their homes and having to leave their village or take shelter on embankments due to natural hazards or exploitation by land-lords or money-lenders. Field workers told me that they feel helpless because they cannot make any structural change in society (compare Goetz, 1996).

PROSHIKA

All PROSHIKA field workers were happy that they had brought some changes to the lives of their clients. They told me what they wanted to achieve in the future. I shall discuss their achievements and then their failures.

I discussed the popularity of the nonformal schools of PROSHIKA. All PROSHIKA field workers told me that the economic situation of most of their clients has improved. They said many of their clients now eat three times a day when it was twice a day before membership. They told me that polygamy and forced divorce by husbands have decreased in their working areas. Due to the voter awareness programme, clients now go to vote in large numbers. All PROSHIKA field workers told me that health awareness among the clients has increased and they now use peat latrines instead of defecating in open places.

All PROSHIKA field workers lamented to me that in many cases the problem of the early marriage of girls has yet to be eliminated. Field workers told me that, when they try to stop early marriage of girls, guardians say that if a girl grows older they may not get a suitable groom and neighbours and relations would say 'why is this man not marrying off his daughter?'. All field workers told me about their helplessness in this situation. The PROSHIKA field workers all told me that more awareness-creation activities are needed otherwise things will remain the same. The best way to change the situation is to convince the people that they themselves have to change their lot. Many longstanding PROSHIKA field workers told me that they think PROSHIKA should re-start their campaign to get state land for the landless. They think that disparity in society may not be eliminated, but it could be reduced by campaigning for land reform and mobilising clients to demand the basic minimum wage which they very often do not get. A major failure identified by the PROSHIKA field workers is again that dowry is still a major problem among their clients.

RDRS Bangladesh

When I asked the field workers of RDRS about their achievements they outlined them as follows.

1. Women have become more aware, they are coming out of their homes in large numbers.
2. Marriages are now registered. Earlier marriages were simply conducted by religious leaders and women could not demand their legal rights during divorce due to lack of documents. The number of forced divorces by husbands has been reduced.
3. More children are going to school.
4. Preparation and use of oral saline has saved thousands of lives from diarrhoea.
5. More households have latrines, which is due to the health-awareness programmes and supply of peat latrines by RDRS.
6. Clients have got many income-generating opportunities and more clients now rear poultry, livestock and do homestead gardening.

RDRS field workers also mentioned to me some of their failures:

1. Dowry is still a major problem among the clients of RDRS. Field workers told me that RDRS has CDECs (Comprehensive Development Education Centres) which offer two hour teaching and discussion programmes. All field workers told me that awareness creation activities are not done very well in those CDECs.
2. Although progress has been made, field workers told me that they feel shocked when many husbands do not allow their wives to join RDRS or attend group meetings saying that RDRS does not give loans quickly enough or that their wives must perform domestic responsibilities.

SCF 'Partner' NGOs

I have already mentioned the clients requests for contraceptives. These field workers are unhappy about their NGOs unwillingness to work on personal maters. SCF 'partner' NGO field workers lamented to me that they feel devastated when they hear about wife-beating among their clients. These NGOs and groups have no policy to act on this. Again, they told me about the problem of the dowry.

The above discussion highlights that field workers know their limitations very well and have good suggestions for the future planning of their NGOs. This knowledge could be a major asset for the NGOs. This understanding has driven me to conduct this research. When I was talking to the field workers they were all telling me that it is the first time they have been asked about their relationship with their clients or superiors and their failures and successes. This once again underlines the necessity of discussion with field workers in any decision-making of the NGOs. The frequent bad relationships between field workers and their superiors can be much reduced by the management of the NGOs, which are increasingly becoming bureaucratic and state-like. This may put the superiority of NGO over state functionaries into question. If poor relationships with field workers continue, the policy-makers of NGOs will have to re-think their ways of working. I shall deal with this in the next Chapter.

Chapter 8

Field Worker Views on NGO Policies and on 'Development'

This Chapter covers two broad themes - the views of the field workers on the activities and policies of their NGOs, and their ideas about 'development'. A third theme, the opinions of field workers on microcredit, cuts across these two and will also be treated here. Sadly, most field workers identify more drawbacks than benefits in the huge complexity of NGO microcredit programmes.

At a conference on 'NGOs in the Global Future' (Birmingham, UK, January 1999) I was repeatedly struck that most discussions were about NGOs, a little on their clients and nothing on their field workers (for details see Edwards *et al.*, 1999). No doubt, NGOs have become prominent, jobs in NGOs have become a profession and research on NGOs attracts fame and funds. This raises the question, why discuss NGOs through the voices of field workers? The answer is, because we need to get these institutions right and for that we need to look back at what NGOs have done and what NGOs should do. This Chapter will be followed by my conclusions, for which I shall try to set the scene here.

Why discuss 'development' with field workers? It is my belief that NGO field workers are the real 'development' workers and have much stronger rights than many others to comment on 'development'. I was trained at Dhaka University and the Asian Institute of Technology in Bangkok to use the word development, but when I went to Durham in 1996 I was advised to write 'development', as the concept had become deeply problematic. Talking to the field workers about their understandings of 'development', I told them this story. All were unhappy at this practice of academics who spend too much time on these trivial issues rather than the main issues. One PROSHIKA field worker said: 'You will never be able to define development, because you do not want it sincerely' (Yaqub Ali, Education Worker, Sakhipur, February, 1998). This statement highlights the academics' problem in talking, thinking and writing about 'development'. This Chapter will therefore discuss what 'development' workers want their clients to achieve.

Field Workers on the Policies of their Own NGOs

Field workers expressed to me their opinions on the policies and working methods of their own NGOs. (Most are women who have little knowledge about the activities of other NGOs so that it would not be feasible to compare their images of other NGOs.) Their opinions will be explored here firstly by NGO, then in all NGOs together.

MCC Bangladesh

Policies towards Staff

1. *Salary cuts as a penalty for failure to get repayment*: I found great dissatisfaction on this issue, usually over microcredit (below). Unlike the three other study NGOs, MCC field workers do not have to worry much about microcredit. But MCC field workers are not free from financial pressure by their organisation. MCC provides materials such as latrine slabs, seed and poultry to its clients but requires eventual payment. Field workers told me that many clients do not repay the money, some intentionally and some due to poverty. Then, on the day the field workers collect their salary they find the accountant has deducted money for the unpaid materials. Mid-level and senior managers of MCC blame the field workers for their laxity. This highlights the problem of switching from relief to service delivery by NGOs, which creates a dependency relationship between the NGOs and their clients. Unfortunately, the ultimate victims of this poor planning are field workers.
2. *Privileges provided to minorities*: In the late 80s and early 90s there were communal, anti-Hindu riots in some parts of Bangladesh. Hindu field workers told me that MCC encouraged them to work in the office to be safe and gave them all possible help. Although MCC could not protect their property, Hindu field workers gratefully acknowledged the generosity of their NGO. I have not heard of such practices in other NGOs.
3. *Work in the rainy season*: All MCC field workers told me about the difficulties in working during the rainy season when farming and other economic activities are interrupted. When monsoon rains are heavy and cause floods, normal life is disrupted and attendance at group meetings is reduced. Field workers also find it difficult to get about in heavy rain and on roads deep in mud, if not under water. After these difficult journeys, field workers face low attendance at group meetings. All field

workers suggested to me that the frequency of group meetings during the rainy season be reduced. Their arguments are very convincing.

4. *Too much paper work*: I have discussed this problem. This is a waste of time and resources. Instead of reporting excessive detail, emphasis should be laid on reporting outcomes, and on reasons for failure and success.

5. *Field Worker/Supervisor/Coordinator Interaction*: The bitter relationship between MCC field workers and their superiors was reviewed in the last Chapter. Field workers may get two different but simultaneous instructions from their supervisor and coordinator respectively, which is a real organisational problem. The disenchantment among field workers with the policy of allowing their supervisors but not themselves to live in town calls for an amicable solution. They want to live in town too. My findings, however, support Jackson (1997a) who wants field workers to live in their working areas so that clients can contact them at any time. Field workers told me that clients cannot afford to go to Maizdee to see them, which seems a valid argument.

Other Policies

1. *Research*: There is general discontent among the field workers of MCC on the research activities of their NGO. Few are well informed about these activities and few find any of them useful for their clients. Most told me that some research is done just to employ researchers. Instead, the field workers (mainly Homestead Development Facilitators) wanted on-farm research and trials to help the clients (farmers). All suggested to me that MCC should pay its clients for on-farm research with new farming methods or agricultural products because poor farmers cannot afford them. Surely, these suggestions from field workers deserve attention from MCC policy makers?

2. *Bargaining with donors*: Some MCC field workers were really bold in criticising donor policies. They think their NGO should motivate their donors to re-orient their emphasis to provide skills training (for example sewing for women, welding for men etc.) for the clients. Field workers told me that the provision of agricultural technology does not help landless or near-landless households which need non-farm activities. One MCC field worker gave me a striking interpretation which represents a level of mistrust and dissatisfaction, common among Southern 'development' workers towards their Northern donors.

Donors wants us to remain dependent on them for their high-priced industrial goods. They do not want us to industrialize. That's why they give us little funding for non-farm activities (Shamsul Alam, Comanyganz, March, 1998).

Service Delivery

Soybeans: Soybeans are a major activity of the MCC Agriculture Programme and MCC has been playing a key role in popularising soybean among its clients in the Noakhali-Lakshmipur area. No doubt, soybeans are a very useful cash crop. They can be used as a nutritious vegetable or a raw material for oil, poultry or fish feed etc. Some field workers told me that ash from soybean plants can be used as washing powder. Soybeans, however, have to compete with paddy rice and other cash crops, mainly ground nuts in the Noakhali-Lakshmipur area.

A major problem of soybeans compared to other cash crops is the limited demand as there is very limited use of soybeans in Bangladesh. Field workers and clients told me that a major problem with soybeans is their price fluctuation, although MCC does buy some seed from its clients. The major buyers of soybeans are poultry firms and fish feed firms. Soybean traders in the local markets told me that, there is insufficient production for a soybean oil factory to be viable in Bangladesh. No doubt, the amount of land that has come under soybean cultivation can be attributed to the hard work of the field workers of MCC rather than to market demand.

Senior managers of MCC told me that NGOs like MCC alone cannot popularise the soybean. It is the state who can do it. According to them, foreign-aided projects to popularise soybeans have failed to yield any results. The administrator of MCC's Agriculture Programme told me that soybeans cannot be made popular without further research and an extension programme from the state. According to him, the price of one kg of soybeans was Tk 13-14 in Bangladesh and Tk 10-11 in the USA, so there is very little room for expanding the market for soybeans at the present level of acreage and production costs (Derek De Silva, Maizdee, March, 1998).

PROSHIKA

Staff Policies

Medical allowance for field workers: This was discussed in detail in Chapter Three. All field workers, and even some mid-level managers, suggested that PROSHIKA should give the allowance without receipts, as to state employees.

Service Delivery

1. *Organic agriculture*: PROSHIKA has an organic agriculture programme which motivates and trains clients for organic farming. All field workers, particularly agriculturists, involved in this programme were unhappy with the whole approach. They do not find it logical to encourage farmers to use cowdung or organic manure instead of chemical fertilisers, because this results in falling production. They say it takes at least three years to get the yields from organic farming equal to that from chemical fertilisers, and poor farmers cannot wait this long. They told me that what actually happens is that farmers use chemical fertilisers but prepare organic fertilisers as instructed by the field workers to show that they have been using them. In their view, dealers and traders in chemical fertilisers are more efficient in selling their products than PROSHIKA is in persuading farmers, so that PROSHIKA alone cannot popularise organic agriculture. They suggested that PROSHIKA should help or provide loans to clients to produce organic fertiliser as they do for producing latrine slabs. They also suggested that PROSHIKA should promote organic agriculture through publications, leaflets and broadcast media, both at local and national levels.
2. *Nonformal schools*: All PROSHIKA field workers were unanimous about the necessity of nonformal schools, and most were happy with the standard of education in them. Some told me that there should be a provision for giving prizes to outstanding students. They said that since there is a system of giving prizes to outstanding students in formal state and private primary schools this frustrates many children in nonformal schools. Field workers told me that, if PROSHIKA cannot afford to provide prizes, groups should come forward for this. This is an appealing proposal.

Other Policies

Comparison between the Grameen Bank (GB) and PROSHIKA: Sohel Chowdhury, Development Worker of PROSHIKA, a field worker who worked for the Bank for about a year, suggested a convincing comparison between it and PROSHIKA (February, 1998). Although the Bank is not an NGO but a specialised bank (Sinha and Matin, 1998), still it is more like an

NGO than a commercial bank. Above all, its main activity is microcredit, which is also the major activity of most NGOs in Bangladesh. (In my view, most NGOs are behaving like microlending agencies like the Grameen Bank.)

Differences between the Grameen Bank and PROSHIKA Identified by Sohel Chowdhury

He thinks the Bank's training programme is more applied than PROSHIKA's. Bank field workers get training at the training centre and in the field on group formation, loan operation and book keeping. After that they have to sit for examinations and be evaluated by their superiors in the field. Only after successful completion of training are field workers absorbed as regular staff. None of this applies to PROSHIKA; so that the Bank's field workers get much better initial training.

The Bank in his view has a more prominent system of supervision than PROSHIKA and field workers have limited personal independence. (This may have been mainly due to the NGO character of PROSHIKA.) The Bank has few residential training programmes for its clients compared to PROSHIKA. (Again, mainly due to the NGO character of PROSHIKA.) The Bank is more strict over loan repayment than PROSHIKA. Clients have to repay their loans in weekly instalments, compared to monthly in PROSHIKA. All Bank clients are women compared to the membership of both men and women in PROSHIKA. Unlike PROSHIKA, GB does not mobilise its clients on local and national issues. (This is a major difference from NGOs like PROSHIKA.) Any Bank field worker who has a first class or classes in their BA/BSc/BCom/BSc(Agr) or MA/MSc/MCom/MSc(Agr) gets one increment for each first class after their full absorption into the service. No doubt, this works as an incentive for meritorious students to join. I have not heard of any such incentive systems in any NGO in Bangladesh.

RDRS Bangladesh

Staff Policies

1. *Unlike most NGOs in Bangladesh, RDRS has a serious problem between its agriculturist and non-agriculturist staff.* A major complaint of the non-agriculturist field workers and mid-level managers against the agriculturists was that the latter's job performance was not tied to microcredit. Nor were they accountable to the mid-level managers at

thana and district levels but only to their project managers at Rangpur and Dhaka. Most non-agricultural field workers and mid-level managers support projects like poultry firms and ponds for fish culture. Most complained to me that agricultural extension field workers often visit these projects and, if they find any problem like the deaths of birds, cattle or fish, report it immediately to their own agriculturist senior manager, not to the non-agriculturist managers. Ultimately the non-agriculturist field workers and mid-level managers have to bear the brunt of these complaints. The non-agriculturist field workers and mid-level managers told me that they are always very busy with microcredit and are not specialised agriculturists. How can they find the necessary time and skill? They also asked me, if the agriculturist field workers are responsible for problems in the agriculturists' projects to Thana and District Managers, why are they needed?

The agriculturist field workers blamed their counterparts for laxity and told me that due to their high workloads, they cannot give enough time to each project. They claimed that the non-agriculturist field workers' and mid-level managers' claim of heavy workloads is merely a pretext (which I find ridiculous). When I was leaving Kurigram (December, 1997) a high-level meeting of RDRS decided to put the agriculturist field workers under the supervision of thana and district level managers. This seems a good decision, but the mistrust between these two groups will take a long time to rectify, as an example will illustrate.

Shafiqul Islam, Agricultural Extension Worker, Ulipur: He was transferred to Ulipur in May 1997. After arriving, he needed accommodation but there was no room in the agricultural field worker's mess (dormitories). He did not want to join the mess of non-agriculturist field workers and they did not want to take him. After seeing his problem, the Thana Manager asked the non-agriculturist field workers to take him in their mess temporarily. The night Mr. Shafiq joined the mess, he was very tired and the next morning he did not wake up on time to get to his office. His colleagues in the mess did not bother to wake him. Mr. Shafiq was very annoyed with his colleagues. Next morning, he was telling his agriculturist colleagues that he was living with some animals who did not bother to wake him up in the morning. One non-agriculturist field worker overheard him from the next room. The non-agriculturist field workers met and forced Mr. Shafiq to leave the mess on the same day (Ulipur, December, 1997).

2. *Paying deposits to field workers*: All RDRS field workers have to deposit Tk 10,000 to become regular staff. The problem arises when field workers leave the job and try to get their money back. All field workers told me that some RDRS managers dilly dally in repayment, so that field workers have to come to the office and meet the relevant staff month after month, which involves costs and wastage of time. Some field workers told me that some of their colleagues had to spend 20 to 40 per cent of their deposits to collect the money. It seems that RDRS should take firm action in reducing the transaction costs of their staff in collecting the deposit.

Service Delivery

1. *Legal aid*: RDRS has a LESS (Legal Education and Support Service) programme for its clients. RDRS usually helps its women clients to secure their legal rights on domestic violence, maintenance, *mohrana* (the money to which women are entitled from their husbands after divorce), polygamy etc. All field workers and some mid-level managers told me that there is rampant corruption in the lower courts in Bangladesh. Therefore, although RDRS pays fees to the lawyers, the field workers cannot pay bribes to the 'other people' in the courts while their opponents can, so many cases fail. They told me that this disadvantage renders their efforts toothless. But in my view, RDRS should continue this programme because the legal rights of women cannot be realised without enforcing them by law.
2. *Tailoring training*: All field workers think RDRS should continue this programme, and if possible expand it.

Other Policies

Wheat sale by guardians: The guardians (all women) of the RDRS Roadside Plantation Programme are paid in wheat (4.5 kg/day) as the programme is run on a partnership between RDRS and the World Food Programme (WFP). On the day when the guardians collect the wheat, traders wait just outside the RDRS office to buy wheat.[1] Immediately after collecting the wheat, guardians sell it to the traders at Tk 8/kg while the market price is Tk 10/kg. There are several reasons why these women sell the wheat at a lower price. *Firstly*, these traders are at the same time money lenders. Due to poverty, most guardians borrow money from these traders

---

[1] My own field observation.

on condition that the money will be repaid in wheat at Tk 8/Kg whatever the market price. *Secondly*, women do not like to go to the local market, move around and bargain with the buyers, so they feel it more convenient to sell it immediately to anyone (even if they have not borrowed money from the traders) and go home. *Thirdly*, many field workers, mid-level workers and guardians told me that on the relevant day all the traders unite to lower their buying price. It may be that if the guardians were paid in cash instead of wheat this could solve the problems (for a change in EU policy see *The Daily Star*, 1999b).

SCF 'Partners'

The area of activity of the 'partners' of SCF is limited as these are small, new NGOs, struggling for survival. (I have discussed why I prefer to refer to SCF (UK) as a donor instead of 'partner' in Chapter Three). The Revolving Loan Fund (RLF) provided by SCF (UK) and the hard work of the field workers have kept the NGOs alive. The total staff salaries as percentage of the total income of the three 'partner' NGOs are- Sakaley Kori, 71 per cent; CWDS, 77 per cent; Upoma, 53 per cent. The condition of the NGOs is therefore precarious because they must maintain a high loan repayment rate to pay their staff salaries, not including office maintenance, transport and other costs. (For the problems of microcredit programmes, see below). The long-term sustainability of these 'partner' NGOs remains in question.

Problems Common to All The Study NGOs

Staff Policies

1. *Promotion and transfer of field workers*: The general dissatisfaction among the MCC field workers on the recruitment and promotion policies of their NGO has been reviewed. There is a lack of transparency in recruitment policies and field workers should be informed by advertisement of opportunities at all levels. In PROSHIKA and RDRS internal memoranda are circulated when recruiting staff, but circulation is often too late, particularly in RDRS. The MCC system of posting field workers to distant districts when promoted requires thorough consideration. Except for the SCF 'partners', field workers of all the study NGOs complained to me about the irregularities in promotion and transfer in their NGOs. This clearly works as a major disincentive for the field workers.

2. *Corruption*: For corruption in some NGOs in Bangladesh see Chapter Six and the case study of Mr. Rahim. RDRS and MCC field workers complained to me about corruption in the recruitment of new field workers in their NGOs.

3. *Discrimination against long-service field workers*: All field workers told me that NGOs do not like long-service field workers to continue in their jobs as field workers. So, field workers who are not promoted are either made redundant, or through unwarranted transfers are compelled to leave their jobs. They gave me two main reasons. *Firstly*, long-service field workers know the ins and outs of their NGOs, about which they can tell new field workers, and some field workers become assertive after working for a long period. *Secondly*, as I have mentioned earlier, NGOs can recruit three new field workers at the cost of one long-service field worker's salary. This is simply the exploitation of youth by NGOs.

4. *Bicycle/Rickshaw*: Among my study NGOs, only MCC and SCF 'partners' do not require their women field workers to ride a bicycle or motor cycle. Women field workers of MCC get the rickshaw fare if they travel more than one km. Mid-level women managers of MCC usually get a jeep or other vehicle to travel to their working areas. I have found a general sense of satisfaction with these policies among the field workers of MCC. In my view, this escape from the bicycle has been possible mainly due to the missionary nature of this NGO.

Service Delivery

Marketing: I asked field workers of my study NGOs about marketing. Interestingly, field workers were unanimous about marketing in the sense that they were against marketing by NGOs because they say the NGOs may exploit their clients, in their view, as BRAC does. Instead, they preferred providing market information and training on marketing to clients.

Field workers also told me about the exploitation of many women clients by their husbands. In many cases, husbands report to their wives low prices of the products they have sold in the market and embezzle the money. Due to purdah it is not possible for women to go to market. When women sell their products at home to the wholesalers, very often they are exploited due to ignorance and helplessness. All these realities represent the limitations of the NGOs in bringing changes to the lives of their women clients.

Other Policies

1. *Top-heavy administration/bureaucratisation*: Except SCF 'partners', the study NGOs have become bureaucratic (compare Onyx and Maclean, 1996; Edwards and Hulme, 1996; Fowler, 1997). The scaling up of NGOs appears to have both pros and cons. On the negative side, the rapid growth of NGOs (in terms of numbers of clients, staff etc.) has detached them from the poor. There are two positive aspects of scaling up: increased prominence of NGOs at the local and national level and the professionalisation which has compelled NGOs to introduce formal staff management policies on promotion, transfer and other fringe benefits. The presence of such policies in MCC, the international NGO in this project, may mainly be due to its missionary nature.

   MCC field workers told me that MCC should start adult education programmes, which help their clients immensely. They also wanted to continue the present children's education programme and to motivate clients to send their children to school. Some told me that MCC should expand its area of activity to other districts to benefit the farmers there. When I asked about resources for the increased area of work and number of clients field workers said that it could be achieved by reducing the number of mid-level and senior managers and the expenditure on research, most of which they find useless. I raised this issue with a senior manager of MCC, who agreed (interview with Derek De Silva, Maizdee, March, 1998). In PROSHIKA most field workers called for more decentralisation of power in loan disbursement. At present an Area Coordinator can sanction loans to a maximum amount of Tk 100,000; a Zonal Coordinator can sanction up to Tk 150,000 and a Principal Project Co-ordinator up to Tk 300,000. A loan for above Tk 300,000 requires permission from the Executive Director. Most field workers think that the Area Coordinator's power should be increased, since more senior people are not always available. This hampers the economic activities and members become unhappy when they do not get the loan at the right time (compare Avina, 1993).

   All field workers, and even some mid-level managers of RDRS, complained to me of the top-heavy staff structure of their NGO. Many field workers of RDRS do not feel the necessity of the next level up and say they can do the work of that level and could be directly accountable to the Thana Managers. Similarly, many managers said that the work of the Credit Manager could be done by an accountant. Field workers and mid-level managers told me that with the recent obsession with microcredit, most subject-matter specialist staff of RDRS, like agriculturists or fishery experts, have become useless. These staff could be reduced to lessen the expenditure of RDRS and enhance its smooth functioning. Some field workers told me that health trainers are unnecessary because they teach the same subjects as other trainers. It seems likely that RDRS needs a re-organisation of its staff structure for both reduction of expenditure and effective functioning.

2. *NGO involvement in politics*: This is an important issue, discussed in Chapter Two. I talked about this with field workers, mid-level and senior managers of the study NGOs. In MCC, mid-level and senior managers are fully against any political role for their NGO. They think that NGOs should always remain non-political, their duty being to provide services to clients. They were critical of the political role of some NGOs. In addition, field workers said that they are already accused of Christian evangelism and have no wish to get embroiled in further problems. Field workers told me that during the last Union Parishad elections, MCC did not give them a holiday (like most other NGOs. Some said that working on election day meant they were accused of working for particular candidates by asking clients to attend meetings or lessons.

PROSHIKA is one of the leading NGOs in Bangladesh to have taken an active role in several political events (ousting the military ruler, movement for holding elections under a neutral caretaker administration etc. For a detailed discussion on the NGOs in politics in the South see Clarke (1998b). PROSHIKA has a voter-awareness programme and encourages its clients to stand in local elections. Field workers are rather sceptical about this political role of their NGO. Most told me that PROSHIKA should not ask its clients to decide to stand in elections or take part in demonstrations for particular parties. Some told me that they were directed by their superiors to mobilise their clients to join specific demonstrations, but see this as a step in the wrong direction. They think that PROSHIKA should not ask its clients to take a party political role: at most, it should make its clients aware of their citizens' rights. Field workers questioned the whole approach of self-sustained group development by PROSHIKA as it asks its clients to take part in an election or demonstration on behalf of a particular party. The latter is clearly a party political role which is illegal for an NGO since they are registered as non-political and non-profit organisations.

All the field workers interviewed, like those of PROSHIKA (above), are against the active involvement of NGOs in party politics. They think at most NGOs should make their clients aware of their voting rights. They think, if any client wants to run in elections at the local or national level, his/her NGO should encourage them in order to make the voices of the landless and poor heard at these levels, whatever their party (local elections in Bangladesh are formally non-political).

3. *Group sustainability*: A major problem identified by all field workers is the sustainability of groups. We have seen in Chapter Six how MCC field workers find it difficult to give less time to old groups and complain of the paucity of educated clients, leadership problems in groups etc. Although I was critical of the field workers for not being able to develop sustainable groups, I have to agree with them that NGO groups are set up by field workers, not clients. Field workers persuade their clients to join groups to get materials and services, which is why clients join NGOs. As we have seen, few NGOs in Bangladesh are grass-root organisations (GROs). With the explosion of NGOs and of their influence, field workers told me that there is very little room for the development of GROs in Bangladesh. All MCC field workers complained to me that most of their clients' incomes will be affected if MCC stopped serving them, again bringing into question claims of sustainable 'development' by NGOs. Overall, it is hardly surprising to see unsustainable NGO groups.

The monthly savings of PROSHIKA groups are deposited with the cashier of the group or somebody assigned by the group. When someone embezzles the savings, this creates conflict in the group and very often they split. Recently PROSHIKA has introduced a system where the savings money is deposited in a bank account which cannot be operated without the consent of the relevant field workers. Field workers told me that the new system has solved an old problem but created a new one. They said if groups cannot operate and maintain their own savings the whole effort by NGOs to promote self-reliant 'development', empowerment and group solidarity come into question.

All the field workers interviewed told me that a major objective of all 'development' work should be to develop sustainable groups of clients. They told me that groups should be able run themselves rather than being dependent on their NGOs. They told me that the whole approach should be to develop leadership, democratic management and autonomy. This problem of group sustainability sheds light on the NGO agenda in Bangladesh where they are serving as the new patrons of their clients.

4. *Partnership*: I have discussed the 'partnership' between SCF (UK) and its 'partner' NGOs in Chapter Three. But I repeat, partnership is a new fashion in the NGO arena in Bangladesh. Apart from the SCF

'partners', the other three study NGOs are involved in different kinds of 'partnership' with other NGOs and donors.

MCC started its 'partnership' programme in 1991. The senior and mid-level managers told me that they thought MCC's technical expertise in agriculture should be extended to other, small NGOs and their clients. Cost-saving, i.e. doing the work through other national NGOs, was another goal in moving into 'partnership' (Derek De Silva, Maizdee, March, 1998).

Field workers of MCC gave me a rather different version of this programme. Although MCC provides agricultural training to the field workers of their 'partner' NGOs, the major activity of the 'partners' is microcredit. MCC field workers still told me that their 'partners' still realise that training in and knowledge of agriculture helps them increase their repayment rates. One advantage of being a 'partner' of MCC is that the 'partner' NGOs can become members of different forums, and involvement with international NGOs helps them get funds from other donors. The advantage for the field workers of 'partner' NGOs is that the training they get from MCC helps them to get jobs in better NGOs. MCC field workers gave me an interesting reason why donors prefer to go for 'partnership', which is economic. Let me elaborate. An MCC field worker on average gets a salary of Tk 5000/month and serves 120 clients. So, his/her cost per client is approximately Tk 42/month. A 'partner' NGO field worker on average gets a salary of Tk 2500/month and serves 200 clients. So, his/her cost per a client is approximately Tk 13/month or 30 per cent of the cost to MCC. The same reason may have driven SCF (UK) to go for 'partnership'.

Just next to the Sakhipur office, PROSHIKA has homestead gardening demonstration plots from a 'partnership' programme with Helen Keller International (HKI). It may be mainly due to its large size that PROSHIKA has more 'partnerships' with smaller NGOs than with donors.

RDRS has a 'partnership' programme with several donors. Like PROSHIKA, RDRS has a 'partnership' programme with HKI. It also has a 'partnership' with the Agricultural Support Services Programme (ASSP) of the Japanese Government for its sewing and tailoring training programmes and with the World Food Programme for plantation and fishpond development programmes. It seems that in Bangladesh the relationships are led by the financial constraints of the Northern NGOs, not by the intention to build a partnership. In other words, it is more donorship than 'partnership'.

5. *Competition among NGOs*: All field workers told me about this problem. There is strong competition among the NGOs in Bangladesh for clients which leads to high client turnover (compare Ebdon, 1995). I have already discussed the competition between SCF 'partners' and SDS in Chapter Three. It seems that SDS will always try to hamper the smooth functioning of the new 'women-run' NGOs, and I warned the senior and mid-level managers of SCF (UK) when I met them. Another result of this competition among the NGOs for clients is the presence of members of different NGOs in one household. This results in the repayment of a loan for one NGO member (say wife) by another member (say husband). Field workers told me that they turn a blind eye to it since they are worried about disbursement and repayment. But in the long run this has disastrous effects, they cautioned me. Another result of this competition among NGOs is the paucity of committed clients for field workers.

In my study NGOs, another reason for high client turnover is the easy availability of credit in NGOs like ASA, BRAC and organisations like Grameen Bank. Some RDRS field workers told me that their client turn over was as high as 25 per cent in some areas. Another reason for this is the availability of services like Vulnerable Group Development cards which enable the clients to get free wheat from other NGOs/GOs. I have heard this allegation mainly against BRAC. Field workers told me that BRAC, Grameen Bank and ASA groups usually have five members and loans are given individually. In those NGOs and organisations, after one member repays a loan, other members get loans easily. Field workers told me that in 20-member PROSHIKA or RDRS groups all loans are given to the groups, so the whole group suffers when one or two members do not repay. They told me that after getting frustrated by the non-repayment of one or two members, many PROSHIKA and RDRS clients leave their NGO because they do not get a new loan even if have they repaid in time.

This competition among the NGO clients points to two key issues:

i) The paucity of co-ordination among NGOs at the field-level.
ii) Instead of trying to reach the vast majority of the poor, NGOs are competing to enrol the same person.

Field Workers' and Clients' Evaluations of Microcredit

In this section I shall discuss the evaluations of field workers, clients and mid-level managers of NGO microcredit programmes. Since my research is on the field workers, the opinions of field workers will dominate. Sadly, field workers highlighted to me the problems of microcredit and told me that microcredit is taking away most of their time and energy, so they have little time left for other 'development' work like education, health awareness or training. Field workers told me that microcredit should be accompanied by education, training and health programmes, not only for high repayment but for the people's needs. They told me that due to overemphasis on credit not all clients come to group meetings, which are essential for discussing all issues related to the group and its activities. Instead, only those clients who want to borrow money or repay come to the meetings, as I have witnessed on many occasions with PROSHIKA, RDRS and SCF 'partners'. Even when all members come to the meetings, most of the time is spent on credit. Field workers also told me that due to disbursement and repayment targets set by their NGOs they do not hesitate to abuse their clients which has, in many cases, damaged the field worker-client relationship. To them, this is a major problem for the NGOs in Bangladesh. They also underlined the fact that microcredit has commercialised the activities of NGOs in Bangladesh (compare Uphoff, 1995). For a detailed discussion on the savings and microcredit programmes of PROSHIKA, see Grace *et al.* (1998). (For the field workers' view on microcredit see Ahmad (1999b).) Throughout the book I have discussed many issues related to microcredit, which I shall try to avoid repeating below.

Microcredit Systems

MCC groups work with both group and individual projects. PROSHIKA and RDRS do not give credit to their clients individually but to groups, the groups being responsible for repayment; all SCF 'partner' NGOs have individual loan systems. The interest rates of loans vary between these four NGOs and between projects, but usually range from 10 to 15 per cent per year. There are penalties for late repayment. Interestingly, the SCF 'partners' do not speak of interest on loans but of a 'service charge'. The charging of any kind of interest is strictly forbidden in Islam. Field workers told me that this is one of the most important reasons that they must get acceptance of their activities in a Muslim community.

All the study NGOs have savings systems (weekly or monthly) and in

MCC clients savings are the only source of credit. PROSHIKA and RDRS borrow capital from different sources because, they say, client savings are not adequate for large-scale microlending. A major source for them is PKSF (a quasi-state organisation, in English the Rural Employment Foundation) which lends to NGOs at a four per cent interest rate (PKSF, 1998). The Revolving Loan Fund (RLF) provided by SCF (UK) to its 'partners' is their major source of capital for lending. Clearly, if an NGO can borrow at four per cent and lend at 10-15 per cent, the difference can become an important source of income if the repayment rate is high.

Why Microcredit?

So why have the NGOs become so obsessed with credit? The answer apparently lies in the minds of the donors who think credit is the best tool to combat poverty and also empower the poor. When RDRS started its activities in the north-west region it was mainly a relief agency. At that time there was no other NGOs there as large as RDRS (in terms of number of clients). Most Thana Managers and their field workers complained that after the intrusion of BRAC, the Grameen Bank and ASA (Assistance for Social Advancement) in the late 1980s, RDRS has started to provide credit to its clients, because when these NGOs started disbursing credit, RDRS started to lose clients. Also, many RDRS clients suggested to their staff that RDRS should start giving credit. In that sense, RDRS is very poorly experienced at operating credit programmes compared to the large NGOs of Bangladesh.

Field workers and mid-level managers suggested several reasons why their NGOs have gone for large-scale microcredit programmes. They told me that an important reason for adopting microcredit as a major activity is the effort by the NGOs to become self-reliant (Edwards, 1999; Vivian and Maseko, 1994). The interest from microcredit helps the NGOs pay some of their staff salary and establishment costs and reduces their dependence on donors. For example, Sakhipur Area (where I worked on PROSHIKA) gets 80 per cent of its costs from microcredit and from commercial operations such as renting out deep tube-wells. That is, the poor pay much of their costs.

Microcredit: Where to Invest?

Field workers, mid-level managers and clients told me about their preferences for microcredit. They told me that agriculture is prone to the vagaries of nature, price fluctuations and, above all, the limited land of the clients makes it difficult to make enough profit to repay loans. Clients prefer to use loans for petty trading, shop-keeping, hoarding rice to sell when the price rises, rice-husking (mostly women), making and selling puffed rice (mostly women), fishing, poultry and cattle rearing.

Loans are given strictly for production, not consumption. Field workers and mid-level managers told me that although they record loans for production or business, clients use them in part for consumption such as payment for marriage of a daughter, examination fees for their children, sending sons overseas to work etc. Field workers, managers and clients told me that they know about this misuse of loans but turn a blind eye because *firstly*, clients want money for their immediate needs, and *secondly*, field workers and managers are worried about repayment not the utilisation of credit.

Field workers and mid-level managers described a range of problems with their microcredit programmes, including the problems which they face in implementing the programmes.

Microcredit Repayment Problems

Field workers gave me several reasons for repayment problems (compare Ebdon, 1995; Goetz, 1996; Ackerly, 1995; Montgomery, 1996). I shall first discuss the reasons specific to the NGOs and then those common to all NGOs. MCC, of course, has limited problems with microcredit.

PROSHIKA

PROSHIKA field workers told me that they think their NGO is less strict with clients about repayment than Grameen Bank, BRAC or ASA. They say that the field workers of these organisations do not leave a meeting until full repayment is made, and in some cases mid-level managers rush to the spot to put pressure on (or to bully) the clients to get the money. PROSHIKA field workers told me they should follow the same method to get a high rate of repayment.

Field workers told me that as a rule they have to keep a level of two million taka in outstanding loans. In doing so, they cannot be very selective in giving loans to credit-worthy groups only. So, they have to give credit to some groups whom they know will not repay, which affects their repayment rates.

Moneylending is a profitable business in Bangladesh. Some clients start moneylending from the savings of their group. This creates conflict in the groups and in some cases they split, again affecting repayment.

RDRS Bangladesh

A major problem reported by field workers is the migration (seasonal and permanent) of clients. Many men RDRS clients migrate to work in other districts during planting and harvesting seasons and lean periods. Being absent from home, these clients do not repay and sometimes they do not return or, if they do, fail to pay the huge accumulated loan.

RDRS field workers as a rule have to keep Tk 50,000 outstanding as loans. Although this is a small fraction of the level in PROSHIKA, they claim to have the same problem. Some field workers told me that they give new loans to allow repayment of the earlier loan to fulfil their target of disbursement and repayment. At some stage this system collapses, resulting in default.

SCF 'Partners' NGOs

To keep a good repayment record, group pressure is the main tool (if one member does not repay regularly other members exert pressure). Also, field workers go to the local leaders to exert pressure on the client to repay loans, but when these leaders are Union Parishad chairmen or members or aspirants, they do not like to antagonise their voters when they are asked by the field workers to exert pressure. So, they do not like to put pressure on their constituents which reduces repayment. Field workers from the influential families in the village are in an advantageous position to get good repayment because poor clients feel obliged to abide by the pressure from local leaders. Some clients still default, saying: 'What can you do if I do not repay?' Some clients told me that sometimes field workers abuse clients for not repaying regularly. Some illiterate clients do not understand accounts and unwarranted misunderstandings with the field workers arise.

Microcredit: General Findings

Field workers of PROSHIKA, RDRS and SCF 'partners' agreed on some reasons which affect the repayment of microcredit. These are:

1. Many shrewd people join NGOs and do not repay loans on the calculation that NGOs will not take legal action against defaulters due to time and resource constraints. They also motivate other clients not to repay, giving the same reasons, which affects repayment.
2. There is a lack of skills among the clients to utilise the loan. Field workers told me that due to paucity of appropriate skills and education,

many clients cannot make best use of their loans. This underlines the need for training and education programmes for clients.

Microcredit: Gains for whom?

All field workers told me that microcredit does not help the poorest. Mosley and Hulme (1998) find from world-wide research that the impact of microlending on the recipient household's income tends to increase (at a decreasing rate) as the recipient's income and asset position improve. This relationship can easily be explained in terms of the greater preference of the poor for consumption loans, their greater vulnerability to asset sales forced by adverse income shocks and their limited range of investment opportunities. Lenders, they argue, can either focus their lending on the poorest and accept a relatively low total impact on household income, or alternatively focus on the not-so-poor and achieve higher impact (Mosley and Hulme, 1998). My findings support this view.

Some SCF 'partner' NGO field workers told me that they still take some of the poorest as members of their NGOs if these can afford to save Tk 5/week. At the end of one year's saving they receive two per cent interest (as a bonus) on their savings, which comes to about Tk 260, and can buy a saree. All other field workers (except MCC) told me that, with the overemphasis on microcredit, NGOs are excluding the poorest from their reach.

Hashemi (1998) argues that although microcredit in Bangladesh, through the Grameen Bank, BRAC, PROSHIKA, ASA and other governmental and non-governmental agencies, has succeeded in reaching a quarter of all poor rural households, poverty still persists. One major reason may be the limits of microcredit in effectively targeting all poor women; or more specifically, in leaving out large sections of the hard core poor - the distressed (Hashemi, 1998; Kabeer, 1998; Independent 1998f; Johnson and Rogaly, 1997). For Hulme and Mosley (1996), the main problems worldwide in the practice of microcredit are overemphasis on credit delivery, social exclusion in the delivery system and a professionalisation of management under which incentive structures for staff, such as bonus payments and promotion prospects, favour concentration on groups other than the core of the poor (Hulme and Mosley, 1996). All these features are highly developed in Bangladesh. Khandker and Chowdhury (1996) hold the relatively positive view that, for the targeted credit programmes in Bangladesh, it will take, on average, about five years for poor programme participants to rise above the poverty line and eight years to achieve 'economic graduation' (to stop taking loans from a targeted credit programme). Montgomery *et al.* (1996), however, found little evidence that BRAC's clientele are altering their structural position within the rural economy. They concluded that credit may be both insufficient and inappropriate for alleviating extreme poverty, an opinion which I share.

All field workers told me that microcredit gives economic lift to at best 60 per cent of their clients. The rest have to work very hard to stay in the same place. The very small amount of credit gives the clients very little economic lift. Alongside this nominal economic progress, other aspects of 'development' in their lives have remained largely unsupported. One SCF 'partner' client told me that her husband had lost most of his money that year because of the low price of his products (chilli), so that her eldest son stopped attending school from May 1997 to catch fish to repay his mother's loans. Although there has been an explosion of microlending organisations (Grameen Bank and NGOs) field workers told me that many clients are still dependent on moneylenders for capital or survival. I have also found that many clients still borrow money from the moneylenders. The reason they gave me for this is that NGO loans are inadequate.

Microcredit: Sad Stories

Field workers reported to me several sad incidents due to microcredit programmes. My findings are similar to those of Rahman (1999). One PROSHIKA field worker told me how shocked he felt when the younger son of a client started crying when his favourite goat was being snatched away from him for non-repayment of credit by his mother. Additionally, the goat was pregnant. Field workers told me how embarrassed they feel when women clients start crying in a meeting due to their inability to repay. Also, field workers told me that there are examples of clients leaving their village in the middle of the night when they become bankrupt and are unable to repay a loan.

Microcredit: Deterioration in Field Worker-Client Relationships

All MCC field workers told me that they prefer MCC's present system under which microcredit is funded from clients' savings instead of by borrowing capital from other agencies. They were still worried about the increasing demand from clients for more microcredit and their threats to leave MCC for other NGOs which give microcredit. All field workers of PROSHIKA, RDRS and SCF 'partners' interviewed told me that their relationships with their clients have deteriorated due to overemphasis on microcredit. Field workers, mid-level managers and clients told me that inter-client relationships have also deteriorated due to credit. There are also many incidences of mistreatment of clients by field workers for not repaying loans. Physical attack and the snatching of furniture or valuables for defaulting are not uncommon. In some cases, the pressure to get back the money borrowed has made field workers cruel. As one PROSHIKA field worker said:

When I lend money, I always put pressure on my clients to repay it by whatever means. I tell them that if you die without repaying my loan I will stamp on your grave four times because you have not repaid the money (Atiqul Alam, Education Worker, Sakhipur, February, 1998).

Field workers told me that, due to client illiteracy and poor understanding of accounts, bitterness between clients and field workers over accounts is common. Some field workers told me that in some cases men clients flee their homes on collection days, disrupting collection and creating uproar in villages. Some RDRS clients told me of a saying that two things wake them up in the morning: a) the call of the field workers or group cashier for money, and b) the call of the beggar who could not pay.

Microcredit has also affected relationships between field workers and their immediate superiors. Field workers told me that they are under double pressure from their superiors to disburse targeted loans and get high repayment rates and from clients for pressuring or maltreating them for defaulting.

Another reason for the deterioration of client-field worker relationships is the strict rules for giving loans, such as regular savings or a minimum amount of savings (10 to 20 per cent of the loan). Field workers nevertheless preferred the continuation of this system which they think works like a collateral against the loan.

Some field workers (mostly RDRS) told me that they do not get adequate co-operation from their superiors in putting pressure on clients to repay loans. They told me that some superiors just give lip-service and do not like to take the trouble to go with the field workers to the field to see the real situation. The Thana Manager of Kurigram Sadar told me that whenever he sees a field worker he asks him or her about their current repayment rate and if it is low, asks why (December, 1997). Mid-level managers told me that they understand the grievances of field workers but they are helpless due to pressure from the top.

Microcredit and New Pressures on Field Workers

My findings are similar to the observations of Ebdon (1995), Goetz and Gupta (1994), Ackerly (1995), Montgomery (1996) and Rahman (1999) (compare Onyx and Maclean, 1996). Take for example PROSHIKA (all ranks of local workers):

1. If their annual repayment rate is 90 per cent or above they qualify for an increment and promotion.
2. If their repayment rate is between 80-90 per cent, they qualify for an increment.
3. If their repayment rate falls below 80 per cent their increment is delayed.
4. If their repayment falls below 70 per cent they get a warning letter and, if this continues for more than a year, they are liable for redundancy (Sakhipur, February, 1998).

If the repayment rate of any RDRS field worker goes below 60 per cent, from the following month he/she has no food allowance until the rate reaches 60 per cent again. This highlights how performance in microcredit has become an indicator for the job performance of NGO field workers in Bangladesh (compare Rao and Kelleher 1998 on BRAC). Since microcredit is the only activity of SCF 'partners', their field workers know that they have to perform well to keep their NGOs alive, keep their jobs and pay their own salaries.

Field workers told me that they have been compelled to become cruel to their clients. When I asked them why clients are still so eager to get credit they told me that a major reason for this is their lack of access to banks.

Microcredit: Field Workers on Loan Use by Men

I agree with Hashemi *et al.* (1996) that participation in credit programmes, whatever the degree of credit transfer to other household members, still brings women a range of personal benefits such as access to training, or status in the household. Simon Mollison, then Country Director of SCF (UK), told me the same (Dhaka, April, 1998). Some field workers told me that some women clients threaten their husbands that if they are treated badly they will not get money from the NGO. I do not totally agree with Goetz (1996) and Goetz and Gupta (1994), who seem critical of the present system of lending money to women which is then used by men. Kabeer

(1998) found that violence to women had been reduced as a result of their access to credit, as did Schuler *et al.* (1998). However, the distribution of violence may be changing rather than the overall impact. Kabeer (1998) concludes that microcredit has given women loanees a greater sense of self-worth and improved marital relationships. Field workers told me that this has happened in the case of some of their clients too. But Rahman (1999) found that violence towards women clients of Grameen Bank has escalated.

Some field workers told me that they do not give loans to those women clients whose husbands misuse it and that some women do not take loans when they fear that their husbands will misuse it. But I had bitter experiences in some PROSHIKA and RDRS working areas where field workers advised their men clients or husbands of their clients to form women's groups to get loans which ultimately went to men. Field workers told me that it is easy to work with women, who are easily available and obedient. They also told me that sometimes husbands come to them for their wives' loans, but they compel the husbands to send their wives. Field workers think this supports the status of women. At the same time, field workers confessed to me that when husbands do not give money to repay, women are usually helpless (compare Rahman, 1999; Ackerly, 1995). This seems inevitable in this traditional Muslim society where women cannot work outside or move freely. But some field workers told me that women clients feel ashamed and cry in front of other clients or field workers when their husbands do not give them the money for instalments. When I met Rina Sen Gupta in Dhaka (April, 1998) she was very critical of the NGO effort to lend money to women. She asked, if NGOs wanted to lend money to men, why should women be used as the means? She told me that NGOs should open banks if they want to work on microcredit in its present form. But this would be a wholesale change.

Field Workers on 'Development'

I asked the field workers about 'development'. What do they think about it? How could it be done better? They gave me their views on service delivery by their NGOs and the state, and on the policies of their NGOs.

NGO Working Methods: All field workers told me that the 'development' effort of NGOs should be directed towards two goals: building social awareness and economic uplift (compare Wood, 1994). For building social awareness they suggested organising regular group meetings to discuss social issues like education, health, women's rights etc. They told me that field workers should ensure that all clients come to the meetings and participate in the discussions. They also suggested work in motivation for both men and women. Regarding the empowerment of women they suggested different measures, such as:

1. Creation of legal awareness, among both men and women, on the rights of women.
2. Creation of health and family planning awareness.
3. Provision of skills training to women for income generation.
4. Education of girls and women.
5. Provision of credit, where necessary, for income generating activities.

For economic uplift they suggested continuing the microcredit programme for small business, cottage industries etc. But they were all against giving credit for the sake of credit (i.e. for the NGO's self-reliance) which does not meet the real needs of the clients and results in mis-targeting (not reaching the poorest) or deepening the problems (giving more credit the clients can absorb) (Matin, 1998; Zaman, 1998).

Field workers emphasised to me that both social awareness and economic 'development' efforts should be pursued simultaneously.

Relief and 'Development': Field workers told me that as a disaster-prone country Bangladesh needs relief after disasters. They suggested that relief should be only for the period of the crisis. After relief, long-term 'development' efforts should be undertaken. They underscored the need for NGOs, GOs and everyone to differentiate between relief and 'development'. All told me about the importance of early warning and evacuation of people, livestock and valuables to minimise loss from disasters.

Family Planning: Field workers emphasised the need to create awareness of family planning among their clients. They were very critical of the state systems of giving money or clothes to those who adopt permanent family planning methods, saying it has resulted in domestic problems in many families and in some cases divorce by husbands. In their eyes, family planning should be achieved through awareness creation not by coercion or money. They suggested that contraceptives should be more easily and cheaply available to women.

Education: All field workers were unanimous on the necessity of both child and adult education. They suggested that the state should learn from the success of the nonformal education programmes of NGOs and adopt some of their syllabus and teaching methods. They also underlined the need for more integration between the nonformal and formal education systems, which are now very different.

Regarding adult literacy classes, field workers told me that their timing should be determined by the clients (men and women separately) for their convenience. They suggested that adult literacy classes should have vacations on the basis of seasonal variations in clients' workloads, but cautioned that long breaks affect enthusiasm and sometimes clients forget the lessons. All field workers emphasised the need for continuing reading and writing lessons through both nonformal and adult literacy education.

Microcredit: All field workers told me that microcredit should be accompanied by training, education and health awareness programmes. Some field workers told me that those NGOs which choose to take microcredit as their main activity and become self-reliant from the interest on their credit should turn themselves into nonprofit banks rather than talking about integrated 'development'. To reduce clients' dependency on NGOs for microcredit, some recommended giving loans to the same client no more than 3 times. They said this would force the clients to use their credit for a productive purpose. They also emphatically said that microcredit will not help the poorest of the poor. They need training and education to sell their labour or skills which will help them to get an economic lift.

Targeting by NGOs: All field workers told me that NGOs should target landless, marginal, farming households, or those households who have to sell their labour for at least 30 days in a year. They also advised that destitute women like the divorced or widowed should be targeted. They were unhappy that in implementing microcredit programmes, NGOs are leaving or being discouraged from reaching the poorest. Some field workers asked me, if NGOs do not target them then who should?

GO-NGO and NGO-NGO Co-operation: All field workers emphasised the need for more, and much more effective GO-NGO and NGO-NGO co-operation at both national and field level. This would eliminate the duplication of work and membership of one client in more than one NGO or people from one household in different NGOs. Above all, they said this would help the NGOs reach more clients instead of competing for the same client or area. They also called for reducing the current level of mistrust and lack of respect between GOs and NGOs and their workers. They said both should have the same goals - reducing poverty and creating social awareness (compare Howes, 1997).

Clients' Dependency on NGOs: All field workers agreed that NGOs are creating a relationship of dependency with their clients. NGO field workers are forming groups and clients are joining the NGOs to get services, so it is a dependent relationship. To solve this problem they suggested that NGOs should clearly state to their clients that they will support them for a maximum of five years. So, the NGOs should train their clients in maintaining accounts, project-planning and implementation and maintenance of discipline in a group. Field workers were very unhappy that after they stopped supporting old groups most of them split or became inoperative.

Why NGOs?: I have discussed this issue with the field workers too. Most field workers told me that the reason for the explosion of NGOs is the failure of the state in reaching and serving the poor, donor preference etc. Some field workers told me that NGOs have been working to resist a revolution in rural Bangladesh (compare Senillosa, 1998). They also pointed out that NGOs do not now talk about land reform or a minimum wage although some of them did in the 70s and early 80s. They told me that NGOs talked about these radical issues to divert the radical workers and thinkers, some of whom are now working in NGOs (see Chapter Two). But all field workers told me that NGOs should bargain with their donors to fund those tasks which are really necessary for their clients. This seems a good proposal for NGOs, but difficult to attain given the relationship between donors and NGOs in Bangladesh.

Conclusion

The issues that I have discussed in this Chapter highlight the big gap between what NGOs do and what their field workers think ought to be done. I suspected this and planned my research on that basis. Unfortunately, my suspicion has proved to be true. In my view this Chapter demonstrates the necessity of more discussion with field workers in the policy making, implementation and evaluation of activities of NGOs.

Chapter 9

Conclusion

Throughout this book, most of my information has come from the field workers. This is not just due to the fact that the research is with the field workers. There is another important reason. Field workers should be the movers and shakers of NGOs. They implement the policies of NGOs, yet they are rarely consulted during the making of these policies. Above all, their problems and opinions remain unheard or accounted for (contrary to the advice of Suzuki, 1998; Fowler, 1997). Throughout the book I have tried to present their voices, views and situation. I shall also conclude with them. I believe field workers know the problems best. They are implementors, so they should be the best people to contribute to realistic solutions. I shall also give my own conclusions drawn from discussions with NGO managers, clients and independent experts. After seeing the NGOs from the bottom up, my realisation is still that field workers need to be empowered to make NGOs more effective, more useful to the poor and the disadvantaged.

I have argued that NGOs in Bangladesh are more businesslike than is usually expected of nonprofit organisations and that they cannot be identified as part of civil society. The recent fashion to make civil society a solution to development problems, in my view, tends to further weaken the state in countries like Bangladesh (Chapter Two).

There is little research on the actual workers of nonprofit organisations in the global North except for volunteers (Clary *et al..*, 1992). Very little research appears to exist on the field workers of NGOs in either North or South Asia, and even less in Bangladesh. I have also discussed the NGOs in Bangladesh. I found them donor dependent (both financially and functionally) and some have been politicised (Chapter Two). I found a recent fashion for Northern NGOs to go into 'partnership' with local NGOs rather than donorship. I found that, except for missionary NGOs like MCC (Mennonite Central Committee), most NGOs in Bangladesh are preoccupied with microcredit, which is mainly driven by the NGOs' search for self reliance and for good performance indicators (Chapter Three). NGO management is very different from that recommended in Human Resource Management by Fowler (1997).

The socio-economic backgrounds of NGO field workers in Bangladesh (Chapter Four) proved to be very distinctive. Most come from the rural middle or lower middle class. Most wanted government jobs and fell back on NGO work when they failed to get them. Most of those who by the time they reach the age of 40 have failed to get promoted either leave voluntarily or are made redundant. Field workers of small NGOs try to switch over to large or international NGOs in search of higher job security and salary and benefits. Due to high unemployment in the country, instead of creating a stable, skilled workforce, NGOs often abandon their long-service field workers in order to recruit cheaper, younger field workers. In my view in the name of helping the poor the NGOs exploit the field workers in the prime of their lives, and then discard them. The personal problems of field workers include job insecurity, financial difficulties, and risks associated with accommodation, child-care and children's education. These problems are more severe for women field workers who also undergo others related to their gender (Chapter Five).

The professional problems of these field workers (Chapter Six) include poor or incorrect training, heavy workloads, corrupt and limited promotion opportunities, irregular and undesirable transfers, low status at work and difficult external relationships. Again, there is a gender difference in these problems. Field workers sometimes forgo promotions to avoid family dislocation (compare Schuler and Huber, 1993). Yet, with all these personal and professional problems, field workers are working and keeping their NGOs running. It is the strengths of the field workers which make this possible. This study calls for better utilisation of these strengths by the NGOs. As Mirvis (1992) observes, in the years ahead differences in work climate across the sectors may erode further and so might the quality of employment in the nonprofit sector. The poor relationship between field workers and their superiors could be resolved by better management in the NGOs (Chapter Seven).

I also found that there is a big gap between what NGOs do and what their field workers think ought to be done. I suspected this and therefore planned the research. Unfortunately, my suspicion has proven to be true. This underscores the necessity of more discussion with field workers in policy making, implementation and evaluation of the activities of NGOs.

Explanations

Before a conclusion is drawn, and some recommendations are made, it is important to ask, 'why are field workers of NGOs in Bangladesh treated like this?' There are several reasons:

The Nature of Bangladesh Society

This requires a brief discussion of the history of Bangladesh society. A self-sufficient, village-based agriculture carried on with a 'primitive plough' and bullock power, and handicrafts made by means of simple instruments, was a basic feature of pre-British colonial South Asian society. A village was almost self-sufficient regarding the raw materials needed for its artisan industry. Another characteristic of the village community was that a rigid caste-structure determined the occupation of its members. Since castes were based on the principle of heredity, occupations also became hereditary (Desai, 1948). Conquerors and invaders, Buddhist, Muslim and Christian have been forced to accept its all-pervading strength.

In what is now South Asia, the early Muslim kings absorbed many of the customs of Hindu rulers, just as the Muslims also developed a caste system (Edwards, 1961). Under British rule, there was a broad correlation between caste and class which duplicated the main classes of the pre-colonial, caste-feudal period. Nevertheless, it was only a correlation, not an identity, and in every caste there could be found some individuals who could get education, a little bit of land, and some access to new opportunities. The fact that artisans, and even untouchables, had formal rights to land ownership, to education and to new occupations was connected with the emergence of 'caste' and 'class' as separate structures, separate but highly interconnected (Omvedt, 1989).

Peasant society in Bangladesh is not a tribal community, nor is it wholly caste-ridden (Jahangir, 1976). Possession of land influences the composition of the household, influences the style of life, and pinpoints the status of the person and the homestead in the social organisation of the rural areas. It is significant that non-economic differences of status, honour and privacy are all attributed along class lines. Mere belonging to particular kin branches does not give 'status honour' unless one has the ability to maintain it. The poorer members of a family, because of their unfavourable economic situation, will generally have minimal social relationships with their richer kin.

The economic distance is increasing between classes, but the inequality is legitimated, the dominance veiled, and the stratification obscured by kinship and quasi-kinship formations in which dominance is legitimised through extra-economic, personalised sanctions. A good example is the hierarchy of social behaviour and etiquette. As Jahangir (1976) observed in the drawing room of one member of a village elite (who was also the wealthiest man), there were chairs and benches. As today, chairs were for the gentlefolk, officials and outsiders of rank; benches were for ordinary people. When a peasant came to see him, the host took his seat on a chair and invited the peasant to sit on the bench. On one occasion, when a father and a son came to see him, he asked the son to sit on a chair and the father on the bench. Later, when Jahangir asked him why, he explained that the son, a college student, would one day be a gentleman but the father was a cultivator; he was treating them according to their occupational status. A student already looked outside his family for education and employment; and peasants were more concerned with developing new types of interpersonal relations than merely acting according to already-allotted roles. This elite man did not reject the trend, but wanted others to come to him and to improve their status with him and through him. He believed that those who are arrogant and defy authority should be prevented from enjoying facilities and privileges, such as share-cropping rights, school scholarships, testimonials for passports, relief, etc. (Jahangir, 1976).

So, what is the status of NGO field workers? The fathers of most field workers work or worked in farming, petty trading or ancillary jobs in government or semi-government offices. Field workers are usually young men and women from middle-class rural families with secondary or higher education who could not get into the civil service. Except in PROSHIKA, most field workers have SSC or HSC qualifications.[1] In PROSHIKA they may be higher. There are several reasons for that. One major reason is that MCC (Mennonite Central Committee) and SCF (UK) have a policy not to accept men or women with degrees as field workers. More specifically, they think field workers do not need to be highly educated and that less educated field workers remain more obedient, although it is usually the good field workers who become mid-level or senior managers. NGO jobs in general are not respected in Bangladesh. People prefer jobs which are permanent (government), well-paid (government, business), powerful (government) and/or office-based (government, sometimes business). Unfortunately field workers' jobs do not enjoy any of these attractions. At the same time, they are often suspected of involvement in Christian evangelism. As workers in jobs despised by the middle class, why should they be treated well by society or their organisations?

The devaluation and maltreatment are more extreme in the case of women staff (compare Fowler, 1997). It seems that male dominance has resulted in this maltreatment by the NGOs (run mainly by men superiors and managers). At the same time, women staff and their views are not

---

[1] SSC is 10 years schooling and HSC is 12 years schooling.

respected by their male colleagues. Bangladesh is a male-dominated, largely rural society. Women remain relatively in the background. In the family, the husband and father is seen as the master. He controls property. He represents the household in its dealings with the outside world. He makes contracts, takes decisions.

Is this because Islam is the dominant religion? The answer is not so simple. Re-reading Islamic texts and tradition, many authorities support women's property rights, the end of polygamy and gender disparities in divorce laws and the rethinking of child custody, plus revival of payments for women at marriage. The diversity found in textual analysis is all the more clear in people's empirical practices and beliefs. There is not one Islam – however much fundamentalists would like to claim this – nor even one Bangladeshi Islam which specifies always and everywhere the same practices and beliefs. These are matters of controversy even within a village: the mullah may speak, but whether or not he is listened to depends on other social and political factors. There is a continuing trend for the use of gender imagery in state discourse with an increasing significance of Islam in this discourse. Nevertheless, while debates about the 'true' interpretation of gender in Islam continue, there is no doubt that its political expression in present day Bangladesh acts significantly to curtail women's room for manoeuvre (White, 1992).

Bangladeshi Organisational Cultures

Organisational cultures in Bangladesh are widely characterised by power distance and lack of concern for juniors (compare Hofstede, 1991). In the power distance situation, parent-child inequality is perpetuated by teacher-student inequality, which caters to the need for dependence well established in the student's mind. The educational process is teacher-centred; teachers outline the intellectual paths to be followed. In the workplace situation, superiors and subordinates consider each other as existentially unequal; the hierarchical system is felt to be based on this existential inequality. Organisations centralise power as much as possible in a few hands. Subordinates are expected to be told what to do (Hofstede, 1991). Field workers appear in the international literature as underpaid, under-valued, overworked, and under-appreciated. In Bangladesh there is a sense in which the juniors normally look up to their seniors, recognising them as superiors and themselves as inferior beings. Centuries of caste society then colonial rule then decades of military or quasi-military rule have left the society and its institutions hierarchical. Power remains concentrated in the hands of a few at the top. The dominant organisational culture in

Bangladesh is to deliver orders from the top and look at the results; managers and decision-makers do not bother about how their orders will be implemented. Any success rewards the managers, but failure is the responsibility of the field workers or subordinates (just as water moves downwards).

From the evidence of this research, the relationship between field workers and their superiors not only depends on the overall management of the NGOs, it also varies from person to person. So, why is this relationship so bad in so many NGOs, or in most offices of some NGOs? As we have just seen, a major reason is the legacy of values which reinforce the practices of traditional societies in South Asia that prioritise men, age and class. After independence, decades of military rule precluded the development of democratic institutions and values. Simon Mollison, then Country Director of SCF (UK), said the same thing (Dhaka, April, 1998). Simon also pointed out to me that the present education system, which has changed little over the years, strengthens this value system. Like the colonial and military rulers, Bangladeshi managers prefer to control and dominate and not to listen to juniors or, sometimes, colleagues. One field worker of PROSHIKA told me that he does not like the system of calling each other 'brothers' or 'sisters' in his NGO because it makes juniors arrogant and disobedient. He preferred the hierarchy of government organisations (Delwar Hossain, Economic Worker, Sakhipur, February, 1998). In state offices (including my semi-state university) juniors are expected to address their seniors as 'sir' or 'madam'. This is an example of the precolonial/colonial value-system among the educated people in the country (let alone the uneducated majority).

Another reason which I would like to mention is also related to such values. As NGOs grow, they become more bureaucratic. All mid-level and senior managers say the same. Although the NGOs started out different from the state (which is highly bureaucratic) as they grew, they developed bureaucracy. The problem appears to be in the wider society. People in the NGOs do not come from outside the society, so gradually the initial distinctiveness of NGOs is lost (Wood, 1997).

Indeed, argues Avina (1993) a principle of good rural and social 'development' practice is that hierarchical centre-field relations are inverted, with field officers as facilitators, advisers and consultants. NGOs which have established a strategic and decentralised management system are structurally more capable of seeing the cause of organisational problems and responding accordingly. In many instances, since the events precipitating consolidation are generally evolutionary in nature, they are even able to foresee difficulties before they manifest themselves and make preventive adjustments in time to limit any negative effects. On the other hand, NGOs bereft of a strategic and decentralised approach to management tend to trip blindly from one crisis to another until they learn their lesson or fail. In many cases there is little feed-back from the field to the head office, where major decisions are made. This puts the field workers in trouble when they are simply asked to meet a target or to work as per the project proposal irrespective of local conditions (Avina, 1993). As the NGOs scale up, the need for decentralised decision-making naturally increases, but rarely comes about, in Bangladesh.

The organisational cultures in the study NGOs have both similarities and differences. A major similarity is task orientation which will be discussed below. It should be noted here that SCF 'partner' NGOs are not comparable to the other three types of study NGOs because of their small size and single gender. The hierarchical structure and lower status of women field workers are common in the other three cases. In that sense, the differences between state organisations and NGOs are to some extent becoming blurred.

NGO Culture in Bangladesh

While fieldwork is generally the 'core process' of an NGO, it is often the least understood process within NGOs in Bangladesh. In other words, NGO culture in Bangladesh is characterised by dependence on charismatic leaders and task orientation. Most NGOs in Bangladesh were formed by these leaders and are still controlled by their wishes. So, Human Resource Development/Management (HRD/M) is irrelevant to them. The leaders think that they know how to get the work done and field workers know the strength and power of the leader. These leaders started these NGOs from zero, so they think their experience and contribution to the growth of their NGOs are more important than following HRD/M policies. The leader's power overshadows the desire of the field workers to ask for formal rules on promotion, transfer etc. If there is a rule the leader (always a man) can easily manipulate or ignore it, because he is the owner of the NGO and field workers cannot afford to go to the courts.

These leaders behave like owners of the NGOs (as in companies or corporations). The governing bodies of the NGOs are in the pockets of the leaders. They can rarely assert their power and influence because there is no trade union or scope for discussion. Leaders also have personal links with the donors, politicians, bureaucrats, academics, and sometimes the media. They know the language of the donors and how to satisfy them. The links are created through providing favours in the form of money (through open or shadow consultancy), jobs (to the bureaucrat/journalist/academic after retirement or to his/her relations) and other benefits like funding foreign trips, or honouring them through inviting them to seminars or cultural functions. All these factors make them unchallenged NGO leaders.

Table 9.1   Hofstede's Five Cultural Dimensions adapted to Bangladeshi NGOs

Power distance	High acceptance of unequal distribution of power
Individualism/Collectivism	Emphasis on individual rights and responsibility
Masculinity/Femininity	Dominance of feminine principles such as creativity, caring, negotiation, persuasion
Uncertainty avoidance	Intolerance of unknown situations with reliance on belief systems, institutions and 'truths'
Time orientation	Short-term values dominate, such as spending, stability, reciprocity, saving 'face', and getting quick results

Source: Adapted by the author from Hofstedc (1991).

Another important feature of NGO culture in Bangladesh is task orientation, alongside the hierarchical structure already discussed. Anything that comes from the top is an order, not advice. There is little discussion between senior and junior staff. So, field workers get orders, are not consulted in planning or policy formulation and are asked to meet targets set by the managers to save jobs or qualify for promotions (if any). The relative neglect of field workers by NGO leaders (and even researchers) and of how field workers use their discretion in implementing policy may reflect a tendency of leaders and researchers to assume that implementation is a mechanical process of carrying out orders, and that changing outcomes is a matter of changing structural features of administration. It is interesting that field workers appeal to Northern norms of fairness, say in talking to me, when the 'culture' of their NGO does not include these norms.

The Saturated Labour Market

A major reason why field workers are treated like this could be the saturated labour market for people with these skills in Bangladesh. Although recently Bangladesh has shown some success in reducing population growth, those who will join the labour force by the year 2010 have already been born. At the same time, the poor performance of the economy (a modest growth rate of around five per cent from a low base) seems unlikely to make any significant change in the depressed labour market for such work. NGOs know there will be no paucity of new applicants for the job of field worker, and attach little value to those they have, so why bother about keeping them?

The Moral Imperative - Reduce Poverty as Cheaply as Possible

Clearly, there is a moral imperative for 'aid' agencies and NGOs to reduce poverty as cheaply as possible. It is difficult to recruit good people in any low-status, low-pay occupation - and very difficult to keep them once recruited. Since field workers are merely the deliverers of the services, no money can be wasted in improving the welfare of the field workers if the goal is to help the poor. (Although it is beyond the scope of this research, a comparison between the benefits enjoyed by the managers and field workers would obviously raise the question of how cheap the activities of NGOs are. For a discussion on the lavish living of NGO managers in Bangladesh, see Hashemi (1995)). Donors want their money and resources to be utilised efficiently and reach the poor. This could be a major reason why donors and NGOs are not concerned about the opinions or welfare of the field workers.

NGOs Become Businesses

Since most NGOs in Bangladesh have come to behave like businesses rather than 'voluntary' organisations, the mission of most of these NGOs has changed (with a few exceptions like MCC Bangladesh). The uncertain funding conditions and changing donor priorities (both geographic and sectoral) may be one reason for this. NGOs are trying to be self-reliant both financially and functionally. The recent popularity of microcredit among NGOs is a good example. So, they have become performance oriented and the jobs of NGO workers have become dependent on their performance in service delivery (compare Fowler, 1997). At the same time, the role of NGOs as organisations caring for their clients has been lost, and their staff are even less considered.

NGOs' Policies towards their Field Workers

There are at least three reasons why NGOs should reconsider their present policy of attaching little or no importance to Human Resource Management and should implement the People in Aid Code (2000). *Firstly*, it is important for ensuring justice for their staff, as for employees in any organisation. *Secondly*, it is essential to improve their motivation and keep them motivated. *Thirdly*, to improve the performance of field workers (Fowler, 1997; Pearson, 1991; Sethi and Schuler, 1989). The recommendations of this research are similar to those of the People in Aid Code although independent of it. Unfortunately, most of the Code's principles remain unrecognised in Bangladesh (Table 9.2). Therefore, it seems important to ascertain how far the NGOs and their donors need to go to implement the Code. Why such principles have not already been adopted has just been discussed, and we have seen that NGOs would need to make some big structural changes to implement Principle 1.

The message of Principle 1 of the Code is clear and supported by this research. Field workers are the driving force behind the effectiveness and success of NGOs. It is disappointing to see that welfare organisations like NGOs need to be asked to ensure the welfare of their staff. Where the change agent is him/herself powerless this also frustrates the goal of empowering their clients. NGOs would have to make big changes in their work cultures and organisational cultures to implement Principle 1. Goetz termed the field-level workers as 'kutcha' (raw) bureaucrats, which indicates the contingent, impoverished, ambiguous role of field workers. Most important, however, is the fact that NGO field workers may be in the least desirable positions in their organisations from a career point of view - careers are not made in the field, nor on women's programmes (Goetz, 1996).

Table 9.2   The Level of Implementation of the People in Aid Code among the Study NGOs

Principle	Implementation
1. The people who work for us are integral to our effectiveness and success	Not at all
2. Our human resource policies aim for best practice	Some in MCC
3. Our human resource policies aim to be effective, efficient, fair and transparent	No
4. We consult our field staff when we develop human resource policy	Only in PROSHIKA. Not through consultation but through field workers' repeated demands, often accompanied by sacrifice.[2]
5. Plans and budgets reflect our responsibilities towards our field staff	Not at all
6. We provide appropriate training and support	No
7. We take all reasonable steps to ensure staff security and well-being	Not at all

Transparency in Human Resource Management

Although it would be good to see the great changes necessary to adopt Principle 1 take place, it is now important to make some practical recommendations which NGOs, state and donors could much more easily implement. Although Principle 2 calls for following the best practices for human resource policies, the findings of this research is that HRM policies are absent (in SCF 'partner' NGOs) or poorly implemented (in the three other NGOs). The slightly better situation in MCC (Mennonite Central Committee) seems to be an exception, but there are very few international missionary NGOs in Bangladesh (discussed below). The presence of irregularities in promotion and transfer impedes the smooth functioning of

---

[2] Demands are often met by punishment such as unwanted transfer or even redundancy.

NGOs. Among the study NGOs, only the small ones (the 'partners' of SCF (UK)) are still free from this problem. NGOs should have clear policies on the transfer, posting and promotion of their field workers which should be properly implemented (Principle 2 of People in Aid Code). The experience and sincerity of field workers should be rewarded at all stages through better management of NGOs.

When available, the HRM policies of NGOs are not effective, efficient, fair nor transparent (Principle 3). Policy or no policy, neither are their practices. The irregularities in transfer and promotion reported in this research are burning examples (Chapter Eight). It does not require big changes in the NGOs and their policies to establish a commitment to be fair in dealing with their staff. The paucity of transparency in the NGOs (let alone HRM policies) is truly frustrating.

NGOs seem to be far from consulting their field workers in formulating HRM policies, let alone allowing them trade unions (discussed below). NGOs could pursue good consultative planning for both HRM and service delivery (Principle 4).

Improving Performance and Motivation

Although there is a moral imperative for reducing poverty and illiteracy as cheaply as possible (discussed above), HRM literature, the People in Aid Code (Principle 5) and the findings of this research all imply a call for better financial packages for the field workers. It is also significant that a study by a Southern researcher on Southern NGOs came to the same conclusions as the HRM practices and Code of best practice prepared in the North.

The training of field workers should therefore be turned into a proper use of their time and of the resources of their NGOs (Principle 6 of People in Aid Code; Fowler, 1997). Field workers should be trained to know the values and policies of their NGOs and apply them in their work. The emphasis and quality of training would demonstrate how NGOs value their field workers as a skilled work force and implementers of their goals. Goetz (1996) found that, due to poor training, field workers did not like to target women simply to challenge the inequities in gender relations. Instead, most offered pragmatic reasons, arguing that women were much more tractable group members and more disciplined loan repayers than men. So, they justify programme delivery approaches which rely upon exploiting women's tractability in the interests of programme efficiency, not women's empowerment.

As a matter of policy, NGOs should require their managers to become more sympathetic to the problems of their field workers. This requires training the managers. This study has found that where managers are sympathetic to the field workers it helps create an amicable relationship among staff and produces better output from the field workers (Principle 7 of People in Aid Code; Fowler, 1997).

In the interest of a more stable, committed and motivated workforce, all NGOs should pursue policies which promote staff welfare, like provident funds, insurance, gratuities and medical allowances (both general and special). (Principle 7 of People in Aid Code). These are very important, considering the risk involved in working in very difficult environments, low social status and insecure job conditions. Field workers travel about with money, and ride motor cycles - these are major risks which should be covered by insurance. At the same time, I would recommend that PROSHIKA give its general medical allowance to all staff as a percentage of their basic salary instead of keeping it subject to the submission of medical evidence. The benefits enjoyed by MCC (Mennonite Central Committee) field workers seem to be exceptional and perhaps often hard to replicate. Not all field workers' complaints are readily resolved: I see no solution to the problem of the family dislocation of NGO field workers (compare Principle 7 of People in Aid Code), as field workers should live in the working areas which is more beneficial for their clients. The problems of women field workers, on the other hand, need due attention from all NGOs as they could be reduced (compare Fowler, 1997).

Improving the Performance of Field Workers

1. Groups of clients are the cornerstones of current NGO practice. To make groups more effective and participatory, more emphasis should be laid on regular attendance at, and participation in, group meetings. Field workers should ensure that groups should discuss not only credit but other issues. I recommend the present system of the SCF 'partners' NGOs which makes the poorest clients members by recognising even very small savings, although the members get an annual profit as low as two per cent. To ensure the participation of the poorest, all NGOs should follow the SCF 'partners' NGOs and make savings more flexible, and participation in meetings more important. This research has found that the poorest stay away from the NGOs due to policies of substantial, compulsory saving and their inability to borrow.

2. NGOs should abandon the present craze for 'performance indicators' (compare Fowler, 1997; Rao and Kelleher, 1998 on BRAC). Businesses can use profit-and-loss measures to evaluate the contribution made by virtually every unit in the company. Their organisational purpose is clear and unambiguous as compared to the purposes of nonprofit organisations. Nonprofit managers may also move their organisations towards precise, measurable objectives and thus reduce uncertainty (Cyert, 1988; Mason, 1996). But this results in some assault on the ideals and mores of the staff and the NGOs. Mason (1996) and Palmer and Hoe (1997) think that the best things are, indeed, often unmeasurable (compare Onyx and Maclean, 1996) and that performance ratings should be much broader. The NGOs should abandon their short-sighted view of 'development' and seek long-term changes in the lives of their clients (compare Edwards, 1999). Field workers should therefore be evaluated not only on the basis of their disbursement and repayment of credit, or on attendance in schools, but on real changes in the lives of their clients (Principle 3 of People in Aid Code). In other words, evaluation should not only be based on figures but on facts too. There are facts behind figures, and NGOs should seek them out. By facts I mean economic prosperity, social awareness etc. For example, to evaluate social awareness, NGOs can assess rates of registered marriages, voting in elections, school enrolment and dropout rates and the reasons for them among the daughters of clients etc. For all these changes, field workers should be adequately trained (Principle 6 of People in Aid Code). But before that, changes are needed in the policies of NGOs which are at present directed towards achieving quick material results (compare Edwards, 1999; discussed below).

A major finding of this research is that NGO field workers are simultaneously social pioneers and 'development' professionals. Although these two roles are not mutually exclusive, the recent 'performance' craze has tilted the balance towards 'professionalism' (compare Slavin, 1988 on staff in the US nonprofits). This has historical links too. In the 1970s and early 80s, NGOs were not so obsessed with performance and microcredit (compare Rao and Kelleher 1998 on BRAC). They were more interested in social mobilisation. Field workers would therefore be able to regard their work as a vocation rather than a profession. Things changed in the 1990s. Goetz (1996) has called the women field workers 'local heroes'. With the recent craze for performance from field workers or the drive to meet indicators, the power and room for manoeuvre for these heroes have been diminishing.

3. Although there are many advantages, this study has found that the general belief that women field workers are better than men for women clients ignores the many difficulties of using women field workers. Once again, this study calls for better training of women field workers for fulfilling the needs of the women clients, particularly the poorest, the majority of whom are widows, abandoned or abused by their husbands. Despite their problems, in this traditional Muslim society, women field workers are still best for women clients, and they should receive support.

4. The findings of this research support Montgomery *et al..* (1996): there is unwarranted competition among the NGOs in the field to target the same client. This should not be allowed to continue when NGOs have so far reached only 10-20 per cent of landless households (Hashemi, 1995). Instead, NGOs should try to reach and serve more clients. There should be co-ordination among the NGOs at the local, regional and national levels to avoid such competition and duplication. ADAB (the Association of Development Agencies in Bangladesh, the NGO umbrella body) can play a vital role here.

5. Most NGOs in Bangladesh have adopted microcredit as their major activity. This study has found that field workers think that microcredit is necessary for the poor but that it is not a panacea. At the same time, while working on microcredit, NGOs are not targeting the poorest who cannot repay loans like the not-so-poor. Even when microcredit is provided to clients it is usually not accompanied by training, education and health awareness programmes.

NGO Client Policies

With regard to the training of their clients, NGOs should re-think their whole strategy. Only training which is useful for clients should be provided. Otherwise, it is a waste of money and resources for both field workers and clients. RDRS (Rangpur Dinajpur Rural Service) should follow the examples of MCC (Mennonite Central Committee) and PROSHIKA in giving food allowances to their clients during training. Above all, training should be followed up and evaluated by the clients. For instance, I appreciate the tailored training by RDRS (Rangpur Dinajpur Rural Service) of its women clients, which should be replicated by other NGOs because it helps them to get employment and generate their own income.

NGO, State and Donor Policies

For NGOs

1. So, what will be the role of NGOs in Bangladesh in the next century? NGOs should simultaneously work for social mobilisation and deliver services to their clients. Social mobilisation includes awareness of the rights of women, the landless, wage labourers, share croppers and all citizens. Service delivery includes provision of education, health awareness, skill-training, credit for income generating trades etc. NGOs should supplement the state by conscientising the poor and providing basic services. The field workers interviewed agree that NGOs should always remain detached from political parties. At best NGOs can make their clients aware of their rights as citizens and promote fair elections which is crucial for Bangladesh. As a first step in this direction, NGOs could promote democratic management of their client groups. Unfortunately, most NGOs in Bangladesh are not managed democratically. The NGOs' umbrella body (which is required to elect its executive committee) is not broad-based. Its membership is often confined to friends and relatives, and elections to the executive committee are often improperly held. This surely frustrates the potential of NGOs as democratic voluntary organisations. Ideally, they should conform to certain standards, and adhere to state regulations (Ahmad, 1999a). Above all, they should take heed of GSS (discussed in Chapter Two). Nevertheless, NGOs cannot function in isolation from the mainstreams of political, economic and social life in the country. Field workers are not allowed to form trade unions. This makes the NGO role in democratic movements ludicrous when NGOs themselves do not like to listen to the voices of their staff.

2. Like Wood (1994) I also found that NGOs in Bangladesh are becoming increasingly bureaucratic. All of them need to learn a lesson from international NGOs like SCF (UK), which was compelled to shed staff due to high operating costs and poor performance (see Edwards, 1999). Here again I find an interesting similarity between NGOs and business. A major challenge for the NGOs in Bangladesh is to remain less bureaucratic, which was one of their advantages over the state. This study has found evidence of bureaucratisation in three out of four examples of NGOs - international, large national and regional (compare Wood, 1994). NGOs should remember that less bureaucracy not only reduces cost but can make the organisation more effective. The originally less hierarchical systems in NGOs, now under threat due to

rapid expansion, can still work as leverage for them over state organisations. The whole NGO agenda was initially driven by their less bureaucratic, more cost-effective organisational culture, which was more effective in reaching and serving the poor compared to the state.

3. A major issue is whether NGOs should be able to promote self-sustained groups of the poor. NGO groups are formed by their field workers, and clients join them to get services. The groups cannot operate independently without the help and supervision of a relevant field worker. So, NGO groups and clients in Bangladesh have become dependent on the services and strengths of the NGOs. This has put the whole question of 'sustainable group development of the poor' into question. NGOs need to promote groups which can operate independently (financially and functionally). This is a real challenge for the NGOs in Bangladesh.

For the State and Donors

1. This study has found evidence of corruption in many NGOs in Bangladesh. State and donors need to be more careful in dealing with NGOs. Some people are doing business in the name of 'development'. So, relevant laws need to be formulated and implemented to make NGOs more transparent and accountable to the state and the people. At the same time, the state must become more transparent and less corrupt (Ahmad, 1999a).

2. There should be more co-ordination between the state, donors and NGOs at the national, divisional and district levels. It is my view from this research that the priorities identified by clients should be accepted as the priorities of the state and donors, which would make national goals more achievable and use resources better. This, of course, would require a complete transformation of national and international practices, which is often discussed but rarely attempted. It is a very long-term hope.

3. I agree with Edwards (1999) who asks the donors not to sacrifice the slow and messy process of institutional development for quick material results; results will come - and will last - if the institutional fabric supports them.

Routes to a Solution?

How may these recommendations be promoted, and what is the likelihood of their happening, given the socio-economic and political situation in Bangladesh?

1. *Will competition between NGOs and in the labour market force a change?* One finding of this research is that field workers switch over from small, low-paying NGOs (like SCF 'partner' NGOs) to large NGOs (like PROSHIKA) and high-paying international NGOs like MCC (Mennonite Central Committee). From my observation, large international NGOs do often set higher standards as employers. In Bangladesh, it is large international NGOs (which have scaled up) which offer better pay and conditions. This, however, proves expensive, and SCF (UK) has responded by scaling down and devolving its work to new local 'partner' NGOs. This also makes opportunities to join an international NGO for better facilities less available. This leaves it to the field workers to negotiate with their own NGOs for improved benefits and working conditions. That has already happened in PROSHIKA. This has in some cases meant a sacrifice, as forced transfer or redundancy may be imposed on field workers for even raising those issues. These all work as disincentives for taking this path. The depressed labour market for field workers (discussed above) makes this situation worse.

2. *Could ADAB address this issue?* ADAB (the Association of Development Agencies in Bangladesh) is the national association of NGO managers and deals with national issues like gender, environment, human rights etc. There is a conflict of interest between ADAB and NGO field workers, similar to that between FBCCI (Federation of Bangladesh Chamber of Commerce and Industries) and trade unions in the manufacturing and service industries, although at least the workers in the latter can form trade unions, and voice their demands. Therefore, it seems less likely that ADAB will do anything for field workers. Because ADAB is composed of NGO managers, it could address the issue, but because their interests are opposed to those of the field workers, it is not likely to do so without specific external pressure or incentives.

3. *Could a path-breaking NGO show a way forward?* Although an NGO could show the way forward, then comes the question, why should it? Given the depressed situation in the labour market and donor reluctance to pay for the welfare of field workers it seems unlikely that a path-

breaking NGO will come forward to make a difference for field workers.

4. *Might NGO trade unions emerge?* Anecdotal evidence from NGOs like PROSHIKA (demands from the field workers in meetings and training sessions) and GSS (sexual harassment of women field workers reported in the press, as discussed in Chapter Two) suggests that there is a possibility that NGO trade unions might emerge. In these situations field workers might assert their human agency. If this happens in one NGO it might show the way for other NGOs, especially if better standards improve the productivity of field workers.

5. *Will donors demand that NGOs improve their HRM?* This is also possible. Bangladesh has two Export Processing Zones (EPZs) where trade unions are not allowed. Recently, under pressure from the US government, trade unions and human rights groups, the Bangladesh government is considering whether to allow limited trade union activities in the EPZs. Similarly, NGO donors could put pressure on NGOs to improve working conditions for field workers of NGOs in Bangladesh. Donors could start with the People in Aid Code and Northern NGOs can urge their Bangladeshi 'partners' to try to implement them. Although it would be a positive development, this would also underline the drawbacks of the whole NGO agenda in Bangladesh which is so donor driven.

Overall, the People in Aid Code needs to be widely distributed and widely read outside the UK as well as within it. It is to be hoped that adoption of the Code will extend rapidly and widely. The Code is evidence of new recognition of, and concern for, field workers. It is an important document and its widespread adoption in the South would transform field workers' lives and effectiveness. No NGO managers (let alone field workers) interviewed in Bangladesh knew about the Code, although People in Aid claims that some NGOs in Bangladesh know about it. It is important to implement it.

By June 1999, nine pilot agencies in the North had completed internal reviews of their progress against the Code indicators. An independent audit was conducted in 2000, still to be published. People in Aid is also considering offering a 'kite mark' for agencies to signal that their commitment to the Code has been audited independently (personal communication, April, 2000). It is encouraging to see that donors (like the Department of International Development (DFID)) have shown interest in the Code. DFID asks those applying to it for emergency funding whether they apply the Code standards, but does not require them to do so. This is important, since donors hold the purse and can create pressure on their recipients. Also some Northern NGOs (like CARE (USA) and World Vision (UK)) are using the Code. 5000 copies have also been distributed in English, Spanish and French (personal communication, April, 2000), unfortunately it has not been distributed in any Southern languages like Bangla. A webpage has been developed, but it has to be remembered that the internet is not accessible to any field workers, only to the managers of some big NGOs in the South. There are no real signs of progress even towards any NNGOs adopting it. It seems that it will take a long time for the field workers of NGOs in Bangladesh to see the Code implemented.

Recommendations for Future Research

1. Research should be carried out with former clients who have ceased to receive help in order to explore clients' dependency of clients on their NGOs.
2. Research could compare field workers of state 'development' organisations and NGOs with regard to motivation, personal and professional problems and opinions on 'development'.
3. Research is needed on the leaders/founders of small NGOs to explore why they have formed their NGOs - whether to do business or assist 'development'.
4. Research could explore the policies and understanding of the Northern donors towards Southern NGOs and their clients, and could make a comparison between the understanding of the Northern donors and realities in the South.
5. Research could be done on the activities of missionary NGOs in the South to inquire into the allegations against them of 'religious conversion'. This would be interesting in the era of a decreasing role for religion in much of the North, and the rise of fundamentalism in some parts of the North and South.

In this book I have aimed to analyse NGOs in Bangladesh in order to argue for changes at all levels - field, management, policy-making, state and donors. I confess my limitations, but have done my best to present the problems and send a message to all. I hope it will not be lost.

Bibliography

Abed, F. H. (1991), 'Extension Services of NGOs: The Approach of BRAC'. Paper presented at the National Seminar on GO-NGO Collaboration in Agricultural Research and Extension. Dhaka, August 4, 1991.

Abed, F. H. and Chowdhury, A. M. R. (1997), 'The Bangladesh Rural Advancement Committee: How BRAC Learned to Meet Rural People's Needs Through Local Action' I Krishna *et al.* (eds). *Reasons for Hope: Instructive Experiences in Rural Development.* Connecticut: Kumarian Press.

Acharya, J. (1997), 'Fake States: Why NGOs Should Not be Civil Society'. *Education Action.* Issue 7. pp. 24-25.

Ackerly, B. A. (1995), 'Testing The Tools of Development: Credit Programmes, Loan Involvement and Women's Empowerment'. *IDS Bulletin.* Vol. 26. No. 3. pp. 56-68.

ADAB (1998), *Directory of NGOs in Bangladesh Ready Reference 1998-99.* Dhaka: ADAB (Association of Development Agencies in Bangladesh).

Adair, J. (1992), *Women in PROSHIKA Kendra (Special Report).* Dhaka: Horizon Pacific International.

Ahmad, M. M. (1991), 'Role of NGOs in The Rural Development of Bangladesh: A Case Study in Income Generation'. Unpublished MSc. Thesis. Department of Geography, University of Dhaka.

Ahmad, M. M. (1993), 'Role of Non-Government Organizations (NGOs) in the Human Resource Development of Rural Bangladesh: A Case Study in Education'. Unpublished MSc. Thesis. Human Settlements Development Programme, School of Environment, Resources and Development, Asian Institute of Technology, Bangkok, Thailand.

Ahmad, M. M. (1999a), 'Non-Governmental Organisations in Bangladesh: An Assessment of Their Legal Status'. Paper presented at the First Asian Third Sector Research Conference. Bangkok, 20-22 November, 1999.

Ahmad, M. M. (1999b), 'Distant voices: Microcredit Finance and the Views of the NGO Field Workers in Bangladesh'. Paper presented at the Conference on NGOs in a Global Future, Birmingham, January, 10-13, 1999.

Ahmed, I. (1997), 'Political System and Political Process in Bangladesh: Consensus on Intolerance and Conflict?'. *The Journal of Social Studies.* No. 78. pp. 71-89.

Ahmed, Y. (1994), 'A Disaster Preparedness Workshop in Pakistan'. *Focus on Gender.* Vol. 2. No. 1. pp. 29-30.

Alam, S. M. (1998), 'NGO Kormi: Durgam Pather Sheshe Jey Dakhey Swapner Bangladesh'. (in Bangla). *Adhuna.* pp. 38-40.

Aminuzzaman, S. (1998), 'NGOs and Grassroot Base Local Government in Bangladesh: A Study of Their Institutional Interactions'. In Hossain, F. and

Myllya, S. *NGOs Under Challenge Dynamics and Drawbacks in Development*. Helsinki: Department for International Development Cooperation/Ministry of Foreign Affairs of Finland. pp. 84-104.

Anheier, H. K. (1990a), 'Private Voluntary Organizations and The Third World: The Case of Africa'. In Anheier, W. and Seibel, H. K (eds). *The Third Sector: Comparative Studies of Nonprofit Organizations*. Berlin: Walter de Gruyter.

Anheier, H. K. (1990b), 'Themes in International Research on The Nonprofit Sector'. *Nonprofit and Voluntary Sector Quarterly*. Vol. 19. pp. 371-191.

Anheier, W. and Seibel, H. K. (1990a), 'Introduction'. In Anheier, W. and Seibel, H. K (eds). *The Third Sector: Comparative Studies of Nonprofit Organizations*. Berlin: Walter de Gruyter.

Anheier, W. and Seibel, H. K. (1990b), 'The Third Sector in Comparative Perspective: Four Propositions' In Anheier, W. and Seibel, H. K (eds). *The Third Sector: Comparative Studies of Nonprofit Organizations*. Berlin: Walter de Gruyter.

Anonymous (1994), 'Bangladesh - Shamiran's Story'. *Alternatives*. Oct. pp. 8-9.

Arnove, R. F. (1980), *Philanthropy and Cultural Imperialism - The Foundations at Home and Abroad*. Boston: G. K. Hall & Co.

Asia Week (1996), 'Activist Power Hits Asia'. www.asiaweek.com. 6 December.

Atack, I. (1999), 'Four Criteria of Development NGO Legitimacy'. *World Development*. Vol. 27. No. 5. pp. 855-854.

Avina, J. (1993) 'The Evolutionary Life Cycles of Non-Governmental Development Organisations'. *Public Administration and Development*. Vol. 13. pp. 453-474.

Bahmueller, C. F. (1997), Untitled. *Civnet*. Vol. 1. No. 1. (From World Wide Web).

Ball, C. and Dunn, L. (1995), *Non-Governmental Organizations: Guidelines for Good Policy and Practice*. London: The Commonwealth Foundation.

Bangladesh Bureau of Statistics (BBS) (1997), *Statistical Pocketbook of Bangladesh 1997*. Dhaka: BBS.

Bangladesh Bureau of Statistics (BBS) (1998), *Statistical Pocketbook of Bangladesh 1998*. Dhaka: BBS.

Bates, R. H. (1981), *Markets and States in Tropical Africa: The Political Basis of Agricultural Policies*. London: University of California Press.

Bauer, R. (1990), 'Nonprofit Organizations in International Perspectives'. In Anheier, W. and Seibel, H. K (eds). *The Third Sector: Comparative Studies of Nonprofit Organizations*. Berlin: Walter de Gruyter.

Bebbington, A. (1997), *Crises and Transitions: Non-Governmental Organisations and Political Economic Change in The Andean Region*. London: ODI/AgREN Network Paper No. 75.

Bebbington, A. J. and Farrington, J. (1993), 'NGO-Government Interaction in Agricultural Technology Development'. In Edwards, M. and Hulme, D. (eds) *Making a Difference: NGOs and Development in a Changing World*. London: Earthscan Publications

Begum, R. (1993), 'Women in Environmental Disasters: The 1991 Cyclone in Bangladesh'. *Focus on Gender*. Vol. 1. No. 1. pp. 34-40.

Bernard, A. *et al.* (eds) (1998), *Civil Society and International Development*. Paris: OECD.

Bhatt, A. (1995), 'Asian NGOs in Development: Their Role and Impact'. In Heyzer, N. *et al. Government-NGO Relations in Asia - Prospects and Challenges for People-Centred Development*. London: Macmillan.

Bhatt, N. (1997), 'Microenterprise Development and The Entrepreneurial Poor: Including The Excluded?'. *Public Administration and Development*. Vol. 17. pp. 371-386.

Blair, H. (1997), 'Donors, Democratisation and Civil Society: Relating Theory to Practice'. In Hulme, D. and Edwards. M. (eds). *NGOs, States and Donors - Too Close for Comfort?* London: Macmillan.

Brazier, C. (1997), 'State of The World Report'. *New Internationalist*. Issue 287. (www.oneworld.org/NI).

Breed, K. (1998), 'Civil Society and Global Governance: Globalisation and The Transformation of Politics'. In Bernard, A. *et al.* (eds). *Civil Society and International Development*. Paris: OECD.

Brenton, M. (1985), *The Voluntary Sector in British Social Services*. London: Longman.

Buvinic, M. and Gupta, G. R. (1997), 'Female Headed Households and Female-Maintained Families: Are They Worth Targeting to Reduce Poverty in Developing Countries?' *Economic Development and Cultural Change*. Vol. 45. No. 2. pp. 259-280.

Cape, N. (1987), 'Papua New Guinea: The Education of an Extension Worker - Thoughts on How an Outsider Can Work in Village Development'. *Reports RRDC Bulletin*. Sept. 1987. pp. 27-32.

Carroll, T. (1992), *Intermediate NGOs: Characteristics of Strong Performers*. Conn: Kumarian Press.

CDRA (1996), *Annual Report 1995/96*. Woodstock: CDRA (Centre for Development Research and Administration).

Chambers, R. (1995), 'NGOs and Development: The Primacy of the Personal'. IDS *Working Paper* 14.

Chowdhury, A. N. (1989), *Let Grassroots Speak: People's Participation Self-help Groups and NGOs in Bangladesh*. Dhaka: UPL.

Christoplos, I. (1998), 'Humanitarianism and Local Service Institutions in Angola'. *Disasters*. Vol. 22. No. 1. pp. 1-20.

CIVICUS (1994), *Citizens Strengthening Global Civil Society*. Washington DC: Civicus.

Clark, J. (1991), *Democratising Development: The Role of Voluntary Organisations*. London: Earthscan.

Clark, J. (1993), 'Policy Influence, Lobbying and Advocacy'. In Edwards, M. and Hulme, D. (eds) *Making a Difference: NGOs and Development in a Changing World*. London: Earthscan Publications.

Clark, J. (1997), 'The State, Popular Participation and The Voluntary Sector'. In Edwards, M. and Hulme, D. (eds). *NGOs, States and Donors - Too Close for Comfort?* London: Macmillan.

Clarke, G. (1998a), *The Politics of NGOs in South-East Asia*. London: Routledge.

Clarke, G. (1998b), 'Non-Governmental Organizations (NGOs) and Politics in The Developing World'. *Political Studies.* XLVI. pp. 36-52.

Clary, E. G. *et al.* (1992), 'Volunteers' Motivations: A Functional Strategy for The Recruitment, Placement, and Retention of Volunteers'. *Nonprofit Management and Leadership.* Vol. 2. No. 4. pp. 333-350.

Colclough, C. and Manor, J. (1991), *States or Markets? Neo-Liberalism and The Development Policy Debate.* Oxford: Clarendon Press.

Connell, D. (1997), 'Participatory Development: An Approach Sensitive to Class and Gender'. *Development in Practice.* Vol. 7. No. 3. pp. 248-259.

Corbridge, S. (1982), 'Urban Bias and Industrialisation: An Appraisal of The Work of Michael Lipton and Terry Byres'. In Harris, J. (ed.). *Rural Development: Theories of Peasant Economy and Agrarian Change.* London: Hutchinson.

Craig, G. and Mayo, M. (eds) (1995), *Community Empowerment - A Reader in Participation and Development.* London: Zed.

Cyert, R. M. (1988), 'The Place of Nonprofit Management Programmes in Higher Education'. In O' Neill, M. and Young, D. R.. *Educating Managers of Nonprofit Organizations.* New York: Praeger.

Dahrendorf, R. (1996), 'Economic Opportunity, Civil Society and Political Liberty'. In Alcantara, C. H. (ed.). *Social Futures, Global Visions.* Oxford: Blackwell/UNRISD.

The Daily Star (1999a), 'Financial Crisis: GSS Closes Operation'. www.dailystarnews.com. 18 April

The Daily Star (1999b), 'EU Prefers Cash-for-work Food Aid May Change its Form'. www.dailystarnews.com. 21 May.

Daniels, A. K. (1988), 'Career Scenarios in Foundation Work'. In O' Neill, M. and Young, D. R., *Educating Managers of Nonprofit Organizations.* New York: Praeger.

Desai, A. R. (1948), *Social Background of Indian Nationalism.* Bombay: OUP.

Devine, J. (1996), 'NGOs: Changing Fashion or Fashioning Change?' CDS Occasional Paper.

De Waal, A. (1997), *Famine Crimes: Politics and the Disaster Relief Industry in Africa.* London: James Currey and Indiana University Press.

Dixon, C. J. and Leach, B. (1977), *Questionnaires and Interviews in Geographical Research.* CATMOG Series No. 18. Norwich: Geo Abstracts.

Drabek, A. G. (1987), 'Development Alternatives: The Challenge for NGOs - An Overview of The Issues'. *World Development.* Vol. 15. Supplement. pp. ix-xv.

Dreze, J. and Sen, A. (1995), *India: Economic Development and Social Opportunity.* Delhi: Oxford University Press.

Eade, D. and Williams, S. (1995), *The Oxfam Handbook of Development and Relief* (3 Vols.). Oxford: Oxfam.

Ebdon, R. (1995), 'NGO Expansion and The Fight to Reach The Poor: Gender Implications of NGO Scaling-up in Bangladesh'. *IDS Bulletin.* Vol. 26. No. 3. pp. 49-55.

The Economist (1998), 'The Other Government in Bangladesh'. July, 25, p.74.

Edwards, M. (1961), *A History of India from Earliest Times to the Present Day.* London: Thames and Hudson.

Edwards, M. (1996), 'NGO Performance - What Breeds Success? A Study of Approaches to Work in South Asia'. SCF-UK *Working Paper* No. 14.

Edwards, M. (1997), 'Organizational Learning in Non-Governmental Organizations: What Have We Learned?'. *Public Administration and Development*. Vol. 17. pp. 235-250.

Edwards, M. (1999), 'NGO Performance - What Breeds Success? New Evidence from South Asia'. *World Development*. Vol. 27. No. 2. pp. 361-374.

Edwards, M. and Hulme, D. (eds) (1993), *Making a Difference - NGOs and Development in a Changing World*. London: Earthscan Publications

Edwards, M. and Hulme, D. (eds) (1993a), 'Scaling-up The Developmental Impact of NGOs: Concepts and Experiences'. In Edwards, M. and Hulme, D. (eds). *Making a Difference - NGOs and Development in a Changing World*. London: Earthscan Publications

Edwards, M. and Hulme, D. (eds) (1995), *Non-Governmental Organisations - Performance and Accountability: Beyond the Magic Bullet*. London: Earthscan Publications.

Edwards, M. and Hulme, D. (eds) (1995a), 'NGO Performance and Accountability: Introduction and Overview'. In Edwards, M. and Hulme, D. (eds) *Non-Governmental Organisations - Performance and Accountability Beyond the Magic Bullet*. London: Earthscan Publications.

Edwards, M. and Hulme, D. (eds) (1995b), 'Beyond The Magic Bullet? Lessons and Conclusions'. In Edwards, M. and Hulme, D. (eds). *Non-Governmental Organisations - Performance and Accountability: Beyond the Magic Bullet*. London: Earthscan Publications.

Edwards, M. and Hulme, D. (1996) 'Too Close for Comfort? The Impact of Official Aid on Nongovernmental Organisations'. *World Development*. Vol. 24. No. 6 pp. 961-973.

Edwards, M. and Hulme, D. (eds) (1997), *NGOs, States and Donors - Too Close for Comfort?* London: Macmillan.

Edwards, M. *et al.* (1999), 'NGOs in a Global Future: Marrying Local Delivery to Worldwide Leverage'. *Public Administration and Development*. Vol. 19. pp. 117-136.

Eickelman, D. F. and Piscatori, J. (1996), *Muslim Politics*. Princeton, New Jersey: Princeton University Press.

Farnworth, E. G. (1991), 'The Inter-American Bank's Interactions with Non-Governmental Organisations'. Paper presented at the Third Consultative Meeting on the Environment. Caracas, June 17-19, 1991.

Farrington, J. and Bebbington, A. (eds) (1993), *Reluctant Partners? Non-Governmental Organisations, the State and Sustainable Agricultural Development*. London: Routledge.

Farrington, J. and Bebbington, A. (eds) (1993a), 'Introduction: Many Roads Lead to NGOs'. In Farrington, J. and Bebbington, A. (eds). *Reluctant Partners? Non-Governmental Organisations, the State and Sustainable Agricultural Development*. London: Routledge.

Farrington, J. and Bebbington, A. (eds) (1993b), 'Concepts for Analyzing NGO-State Relationships'. In Farrington, J. and Bebbington, A. (eds). *Reluctant*

Partners? Non-Governmental Organisations, the State and Sustainable Agricultural Development. London: Routledge.

Farrington, J. and Lewis, D. J. (eds) (1993), *Non-Governmental Organisations and The State in Asia - Rethinking Roles in Sustainable Agricultural Development.* London: Routledge.

Farrington, J. and Lewis, D. J. (eds) (1993a), 'Introduction'. In Farrington, J. and Lewis, D. J. (eds). *Non-Governmental Organisations and The State in Asia - Rethinking Roles in Sustainable Agricultural Development.* London: Routledge.

Farrington, J. and Lewis, D. J. (eds) (1993b), 'Policy Implications: Reconciling Diversity with Generalization'. In Farrington, J. and Lewis, D. J. (eds). *Non-Governmental Organisations and The State in Asia - Rethinking Roles in Sustainable Agricultural Development.* London: Routledge.

Feldman, S. (1997), 'NGOs and Civil Society: (Un)stated Contradictions'. *Annals of The American Academy of Political and Social Science. The Role of NGOs: Charity and Empowerment.* No. 554. pp. 46-65.

Fisher, W. F. (1997), 'Doing Good? The Politics and Antipolitics of NGO Practices'. *Annual Review of Anthropology.* Vol. 26. pp. 439-464.

Fowler, A. (1997), *Striking a Balance A Guide to Enhancing The Effectiveness of Non-Governmental Organizations.* London: Earthscan.

Fowler, A. (1988), 'NGOs in Africa: Achieving Comparative Advantage in Relief and Micro-development'. IDS *Discussion Paper* No. 149.

Frenk, S. F. (1995), 'Re-presenting Voices - What's Wrong with Our Life Histories?'. In Townsend, J. G. *et al.* (1995) (eds). *Women's Voices from The Rainforest.* London: Routledge.

Friedman, J. (1998), 'The New Political Economy of Planning: The Rise of Civil Society'. In Douglas, M. and Fiedman, J. (eds). *Cities for Citizens - Planning and The Rise of Civil Society in a Global Age.* Chichester: John Wiley & Sons.

Friedman, J. (1992), *Empowerment: The Politics of Alternative Development.* Oxford: Basil and Blackwell.

Fyvie, C. and Ager, A. (1999), 'NGOs and Innovation: Organizational Characteristics and Constraints in Development Assistance Work in The Gambia'. *World Development.* Vol. 27. No. 8. pp. 1383-1385.

Garain, S. (1993), 'Training Grassroots Level Workers in Empowering The Rural Poor: The Case of an Indian NGO'. *The Indian Journal of Social Work.* Vol. LIV. No. 3. pp. 381-392.

Giddens, A. (1984), *The Constitution of Society.* Cambridge: Polity Press.

Goetz, A. M. (1995), 'Employment Experiences of Women Development Agents in Rural Credit Programmes in Bangladesh: Working Towards Leadership in Women's Interests'. Sussex: IDS *Working Paper.*

Goetz, A. M. (1996), 'Local Heroes: Patterns of Field Worker Discretion in Implementing GAD Policy in Bangladesh'. IDS Discussion Paper 358.

Goetz, A. M. (1997), 'Managing Organizational Change: The 'Gendered' Organisation of Space and Time'. *Gender and Development.* Vol. 5. No. 1. pp. 17-27.

Goetz, A.M. and Gupta, R. S. (1994) ' Who Takes the Credit? Gender, Power, and Control over Loan Use in Rural Credit Programmes in Bangladesh'. IDS *Working Paper 8.*

Grace, *et al.* (1998), *PROSHIKA Manobik Unnayan Kendra Bangladesh - Report of The Savings Review.* Dhaka (mimeo).

Griffith, G. (1987), 'Missing The Poorest: Participant Identification by a Rural Development Agency in Gujarat'. IDS *Discussion Paper 230.*

Grindle, M. S. (1986), *State and Countryside - Development Policy and Agrarian Politics in Latin America.* Baltimore: Johns Hopkins University Press.

Guardian/WWF-UK (1997), *One Thousand Days - A Special Report on How to Live in The New Millennium.* London: Guardian/WWF-UK.

Hadenius, A. and Uggla, F. (1996), 'Making Civil Society Work, Promoting Democratic Development: What Can States and Donors Do?'. *World Development.* Vol. 24. No. 10. pp. 1621-1639.

Harvey, P. (1998), 'Rehabilitation in Complex Political Emergencies: Is Rebuilding Civil Society The Answer?'. *Disasters.* Vol. 22. No. 3. pp. 200-217.

Hasan, A. (1993), *Scaling-up The OPP's Low-Cost Sanitation Programme.* Karachi: OPP.

Hashemi, S. M. (1992), 'State and NGO Support Networks in Rural Bangladesh: Concepts and Coalitions for Control'. (mimeo) Copenhagen: Centre for Development Research.

Hashemi, S. M. (1995), 'NGO Accountability in Bangladesh: Beneficiaries, Donors and State'. In Edwards, M. and Hulme, D. (eds) *Non-Governmental Organisations - Performance and Accountability: Beyond the Magic Bullet.* London: Earthscan Publications.

Hashemi, S. M. (1998), 'Those Left Behind: A Note on Targeting The Hard-core Poor'. In Wood, G and Sharif, I. A. (eds). *Who Needs Credit - Poverty and Finance in Bangladesh.* London: Zed Books.

Hashemi, S. M. and Hassan, M. (1999), 'Building NGO Legitimacy in Bangladesh: The Contested Domain'. In Lewis, D. (ed.). *International Perspectives on Voluntary Action: Reshaping The Third Sector.* London: Earthscan.

Hashemi, S. M. and Hossain, Z. (1995), 'Evaluation of Knowledge and Skills of Field Level Workers of Health and Family Planning Programmes'. Dhaka: Ministry of Planning, Population and Evaluation Unit, GOB.

Hashemi, S. M., Schuler, S. R. and Riley, A. N. (1996), 'Rural Credit Programmes and Women's Empowerment in Bangladesh' *World Development.* Vol. 24. No. 4. pp. 635-653.

Haverkort, B., Van Der Kemp, J. and Waters-Bayer, A. (eds). (1991), *Joining Farmer's Experiments: Experiences in Participatory Technology Development.* London: IT Publications.

Healy, J. and Robinson, M. (1992), *Democracy, Governance and Economic Policy: Sub-Saharan Africa in Comparative Perspective.* London: ODI.

Heyns, S. (1996), 'Organisational Capacity-Building, and the 'Quick and Dirty Consultant'. *Development in Practice.* Vol. 6. No.1. pp. 54-57.

Hirschmann, D. (1999), 'Development Management Versus Third World Bureaucracies: A Brief History of Conflicting Interests'. *Development and Change*. Vol. 30. pp. 287-305.

Hofstede, G. (1991), *Cultures and Organisations – Software of the Mind*. London: McGraw-Hill.

Holcombe, S. (1995), *Managing to Empower - The Grameen Bank's Experience of Poverty Alleviation*. London: Zed Books.

Holloway, R. (ed.) (1989), *Doing Development - Governments, NGOs and Rural Poor in Asia*. London: Earthscan/CUSO.

Holman, M. (1999), 'The 'Firemen' of Africa Feel The Heat of Scrutiny'. *Financial Times*. August, 19.

Hoque, R. and Siddiquee, N. A. (1998), 'Grassroots Democratisation in Bangladesh: The NGO Experience'. *The Journal of Social Studies*. No. 79. pp. 49-55.

Hossain, F. and Myllyla, S. (1998), 'Introduction - Non-Governmental Organisation: In Search of Theory and Practice' In Hossain, F. and Myllya, S. *NGOs Under Challenge: Dynamics and Drawbacks in Development*. Helsinki: Department for International Development Cooperation/Ministry of Foreign Affairs of Finland. pp. 13-21.

Hossain, M. B., Phillips, J. F. and Haaga, J. G. (1994) ' The Impact of Field Worker Visits on Contraceptive Discontinuation in Two Rural Areas of Bangladesh'. Paper presented at the Annual Meeting of the Population Association of America, Miami, Florida. May 5-7, 1994.

Howes, M. (1997), 'NGOs and The Institutional Development of Membership Organizations: The Evidence From Six Cases'. *Journal of International Development*. Vol. 9. No. 4. pp. 597-604.

Hulme, D. and Mosley, P. (eds) (1996), *Finance Against Poverty*. Vol. 1. London: Routledge.

Hyden, G. (1983), *No Shortcuts to Progress - African Development Management in Perspective*. Berkeley: University of California Press.

IHE (International Health Exchange)/People in Aid (1997), *The Human Face of Aid - A Study of Recruitment by International Relief and Development Organizations in the UK*. London: IHE/People in Aid.

Independent (Dhaka) (1998a), 'Stress on Micro-Credit to Fight Poverty'. www.independent-bangladesh.com. 12 July.

Independent (Dhaka) (1998b), 'Micro-Credit Only Won't Alleviate Poverty: Kibria'. www.independent-bangladesh.com. 9 July.

Independent (Dhaka) (1998c), 'Fake NGO Grabs Tk. 1.5 Crore'. www.independent-bangladesh.com. 2 August.

Independent (Dhaka) (1998d), '18 NGOs Selected to Set Up 270 Primary Schools'. www.independent-bangladesh.com. 3 June.

Independent (Dhaka) (1998e), 'NGOs Should be Accountable for Their Activities: Motia'. www.independent-bangladesh.com. 15 July.

Independent (Dhaka) (1998f), 'Hardcore Poor have little Access to Micro-credits'. www.independent-bangladesh.com. 7 November.

Independent (Dhaka) (1999a), 'ADAB on Verge of Split over Slum Demolition'. www.independent-bangladesh.com. 6 September.

Islam, S. (1995), 'Middle-Income Women in Dhaka City: Gender and Activity Space'. Unpublished MA. Thesis. Department of Geography, University of Durham, UK.

Jackson, C. (1997a), 'Sustainable Development at the Sharp End: Field-Worker Agency in a Participatory Project' *Development in Practice*. Vol. 7. No. 3. pp. 237-247

Jackson, C. (1997b), 'Actor Orientation and Gender Relations at a Participatory Project Interface'. In Goetz, A. M. (ed.). *Breaking In, Speaking Out.* London: Zed Press.

Jahan, R. (1988), 'Hidden Wounds, Visible Scars - Violence Against Women in Bangladesh'. In Agarwal, B. (ed.) *Structures of Patriarchy*. London: Zed Books.

Jahangir, B. K. (1976), 'Differentiation, Polarisation and Confrontation in Rural Bangladesh'. Unpublished Ph.D. Thesis. Department of Anthropology, University of Durham, UK.

Jahangir, B. K. (1997), 'Reformist Agenda for Bangladesh'. *The Journal of Social Studies*. No. 75. pp. 23-32.

Jain, P. S. (1996), 'Managing Credit for the Rural Poor: Lessons from the Grameen Bank'. *World Development*. Vol. 24. No. 1 pp. 79-89.

James, E. (1990), 'Economic Theories of The Nonprofit Sector'. In Anheier, W. and Seibel, H. K (eds). *The Third Sector: Comparative Studies of Nonprofit Organizations*. Berlin: Walter de Gruyter.

Jeavons, T. H. (1994), *When The Bottom Line is Faithfulness - Management of Christian Service Organizations*. Bloomington: Indiana University Press.

Johnson, B. L. C. (1975), *Bangladesh*. London: Heinemann.

Johnson, S. and Rogaly, B. (1997), *Microfinance and Poverty Reduction*. Oxford: Oxfam and ActionAid.

Kabeer, N. (1998), 'Money Can't Buy Me Love? Re-evaluating Gender, Credit and Empowerment in Rural Bangladesh'. IDS *Discussion Paper* 363.

Kabir, N. (1999), 'In a Mess: One of The Better-Known NGOs Faces Charges of Irregularities'. *The Daily Star*. www: dailystarnews.com. 25 May.

Kandil, A. (1997), 'Egypt'. In Salamon, L. M. and Anheier, H. K. (eds). *Defining The Nonprofit Sector A Cross-national Analysis*. Manchester: Manchester University Press.

Khandker, S. R. and Chowdhury, O. H. (1996), *'Targeted Credit Programmes and Rural Poverty in Bangladesh'*. World Bank *Discussion Paper* No. 336. Washington DC: The World Bank.

Kibble, S. (1997), 'Conclusions' In Baranyi, S., Kibble, S., Kohen, A. and O'Neill (eds) *Making Solidarity Effective: Northern Voluntary Organisations Policy Advocacy and The Promotion of Peace in Angola and East Timor*. London: CIIR.

Kirlels, E. (1990), *Role and Situation of Women Development Workers in Non Governmental Organizations in Bangladesh*. Bonn: EZE.

Knapp, M. (1996), 'Are Voluntary Agencies Really More Effective?'. In Billis, D. and Harris, M. (eds). *Voluntary Agencies - Challenges of Organisation and Management*. London: Macmillan.

Korten, D. C. (1987), 'Third Generation NGO Strategies: A Key to People-Centred Development'. *World Development*. Vol. 15. Supplement. pp. 145-159.

Korten, D. C. (1990), *Getting to The 21st Century - Voluntary Action and the Global Agenda*. Conn: Kumarian Press.

Korten, D. C. (1995), 'Steps Toward People-Centred Development: Vision and Strategies'. In Heyzer, N. *et al*. (eds) *Government-NGO Relations in Asia-Prospects and Challenges for People-Centred Development*. London: Macmillan.

Korten, D. C. and Quizon, A. B. (1995), 'Government, NGO and International Agency Cooperation: Whose Agenda?'. In Heyzer, N. *et al*. (eds) *Government-NGO Relations in Asia - Prospects and Challenges for People-Centred Development*. London: Macmillan.

Kothari, S. (1999), 'Inclusive, Just, Plural, Dynamic: Building a 'Civil' Society in The Third World'. *Development in Practice*. Vol. 9. No. 3. pp. 246-259.

Kramer, R. M. (1990), 'Nonprofit Social Service Agencies and The Welfare State: Some Research Considerations'. In Anheier, W. and Seibel, H. K. (eds). *The Third Sector: Comparative Studies of Nonprofit Organizations*. Berlin: Walter de Gruyter.

Kretschmer, K and Elwert, G. (1991), 'The Target Group's Perspective on The Expatriate Volunteers in Benin' (in German). *Afrika Spectrum*. Vol. 26. No. 3. pp. 335-350.

Kruger, A., Schiff, M. and Valders, A. (1991), *The Political Economy of Agricultural Pricing Policy*. Vol. 1. Latin America, Baltimore: Johns Hopkins University Press.

Krut, R. (1997), Globalisation and Civil Society - NGO Influence in International Decision-Making. UNRISD *Discussion Paper* P. 83. Geneva: UNRISD.

Lal, D. (1996), 'Participation, Markets and Democracy'. In Lundahl, M. and Ndulu, B. J. (1996) (eds). *New Directions in Development Economics - Growth, Environmental Concerns and Government in the 1990s*. London: Routledge.

Lawrence *et al*. (1998), *The Blackwell Encyclopedic Dictionary of Human Resource Management*. Oxford: Blackwell.

Leat, D. (1990), 'Voluntary Organizations and Accountability: Theory and Practice'. In Anheier, W. and Seibel, H. K (eds). *The Third Sector: Comparative Studies of Nonprofit Organizations*. Berlin: Walter de Gruyter.

Leat, D. (1996), 'Are Voluntary Organisations Accountable?'. In Billis, D. and Harris, M. (eds). *Voluntary Agencies - Challenges of Organisation and Management*. London: Macmillan.

Legge, K. (1995), *Human Resource Management – Rhetorics and Realities*. London: Macmillan.

Lehman, A. D. (1990), *Democracy and Development in Latin America - Economics, Politics and Religion in The Post-War Period*. Cambridge: Polity Press.

Lehning, P. B. (1998), 'Towards a Multi-Cultural Civil Society: The Role of Social Capital and Democratic Citizenship'. In Bernard, A. *et al.* (eds). *Civil Society and International Development*. Paris: OECD. pp. 27-42.

Lewis, D. J. (1993), 'NGO-Government Interaction in Bangladesh'. In Farrington, J. and Lewis, D. J. (eds) *Non-Governmental Organisations and The State in Asia - Rethinking Roles in Sustainable Agricultural Development*. London: Routledge.

Lewis, D. (1997), 'NGOs, Donors, and The State in Bangladesh'. *Annals of The American Academy of Political and Social Science. The Role of NGOs: Charity and Empowerment*. No. 554. pp. 33-45.

Lewis, D. (n.d.), 'Bridging The Gap? The Parallel Universe of The Non-Profit and Non-Governmental Organisation Research Traditions and The Changing Context of Voluntary Action'. CVO *International Working Paper* No. 1.

Lipton, M. (1977, 1986), *Why Poor People Stay Poor: A Study of Urban Bias in World Development*. London: Temple Smith; Aldershot: Gower.

Long, N. (1988), 'Sociological Perspectives on Agrarian Development and State Intervention'. In Hall, A and Midgley, J. (eds). *Development Policies: Sociological Perspectives*. Manchester: Manchester University Press.

Long, N. (1992), 'Conclusion'. In Long, N and Long, A (eds). *Battlefields of Knowledge - The Interlocking of Theory and Practice in Social Research and Development*. London: Routledge.

Macdonald, L (1995), 'A Mixed Blessing - The NGO Boom in Latin America' *NACLA Report on the Americas*. Vol. xxviii. No. 5. pp. 30-33.

Macdonald, L. (1997), *Supporting Civil Society - The Political Role of Non-Governmental Organizations in Central America*. London: Macmillan.

Macnair, R. (1995), *Room for Improvement: The Management and Support of Relief and Development Workers*. London: ODI.

Malhotra, K. (1997), 'Something Nothing Words: Lessons in Partnership from Southern Experience'. In Hately, L. and Malhotra, K (eds). *Between Rhetoric and Reality: Essays on Partnership in Development*. Ottawa: The North South Institute.

Malkia, M. and Hossain, F. (1998), 'Changing Patterns of Development Co-operation: Conceptualising Non-Governmental Organisations in Development'. In Hossain, F and Myllya, S. (eds). *NGOs Under Challenge Dynamics and Drawbacks in Development*. Helsinki: Department for International Development Cooperation/Ministry of Foreign Affairs of Finland. pp. 22-46.

Manji, F. (1997), 'Collaboration with South: Agents of Aid or Solidarity?'. *Development in Practice*. Vol. 7. No. 2. pp. 175-178.

Marshall, T. F. (1996), 'Can We Define The Voluntary Sector?'. In Billis, D. and Harris, M. (eds). *Voluntary Agencies - Challenges of Organisation and Management*. London: Macmillan.

Martinez, N. R. (1990), 'The Effect of Changes in State Policy and Organisations on Agricultural Research and Extension Links: A Latin American Perspective'. In Kaimowitz, D. (ed.). *Making The Link: Agricultural Research and Technology Transfer Services in Developing Countries.* London: Westview Press.

Maslow, A. (1970), *Motivation and Personality*, 2nd ed., Harper & Row.

Mason, D. E. (1996), *Leading and Managing The Expressive Dimension Harnessing The Hidden Power Source of The Nonprofit Sector.* San Francisco: Jossey-Bass Publications.

Matin, I. (1998), 'Mis-Targeting by The Grameen Bank: A Possible Explanation'. *IDS Bulletin.* Vol. 29. No. 4. pp. 51-58.

McIlwaine, C. (1998), 'Civil Society and Development Geography'. *Progress in Human Geography.* Vol. 22. No. 3. pp. 415-424.

Mennonite Central Committee (MCC). (1998), *Mennonite Central Committee.* Dhaka: MCC

Meyer, C. (1992), 'A Step Back as Donors Shift Institution Building from The Public to The Private'. *World Development.* Vol. 20. No. 8. pp. 1115-1126.

Meyer, C. A. (1997), 'The Political Economy of NGOs and Information Sharing'. *World Development.* Vol. 25. No. 7. pp. 1127-1140.

Mirvis, P. H. (1992), 'The Quality of Employment in The Nonprofit Sector: An Update on Employee Attitudes in Nonprofit Versus Business and Government'. *Nonprofit Management and Leadership.* Vol. 3. No. 1. pp. 23-41.

Mitchell, B. (1989), *Geography and Resource Analysis.* New York: John Wiley.

Mollison, S. (1996), 'An Unplanned Planning Process'. A Briefing Paper Prepared by Save The Children-UK. London: SCF-UK.

Montgomery, R. (1996), 'Disciplining or Protecting The Poor? Avoiding The Social Costs of Peer Pressure in Micro-Credit Schemes'. *Journal of International Development.* Vol. 8. No. 2. pp. 289-305.

Montgomery, R. *et al.* (1996), 'Credit for The Poor in Bangladesh - The BRAC Rural Development Programme and The Government Thana Resource Development and Employment Programme'. In Hulme, D. and Mosley, P. (eds). *Finance Against Poverty.* Vol. 2. London: Routledge.

Moore, B. (1996), *The Social Origins of Dictatorship and Democracy*, Boston: Beacon Press.

Moore, M. (1993), 'Good Government? Introduction'. *IDS Bulletin.* Vol. 24. No. 1. pp. 1-6.

Moore, M. and Stewart, S. (1998), 'Corporate Governance for NGOs?'. *Development in Practice.* Vol. 8. No. 3. pp. 335-342.

Morgan, G. (1997), *Images of Organization.* London: Sage Publications.

Morgan, M. (1990), 'Stretching The Development Dollar: The Potential for Scaling Up'. *Grassroots Development.* Vol. 14. No. 1. pp. 2-12.

Mosley, P., Harrington, J. and Toye, J. (1991), *Aid and Power: The World Bank and Policy Based Lending.* Vol. 1. London: Routledge.

Mosley, P. and Hulme, D. (1998), 'Microenterprise Finance: Is There a Conflict Between Growth and Poverty Alleviation?' *World Development.* Vol. 26. No. 5. pp. 783-790.

Mouzelis, N. (1995), 'Modernity, Late Development and Civil Society'. In Hall, J. A. (ed.). *Civil Society - Theory, History, Comparison.* Cambridge: Polity Press.

Murdoch, J. and Marsden, T. (1995), 'The Spatialization of Politics: Local and National Actor-Spaces in Environmental Conflict'. *Transactions of The Institute of British Geographers.* NS Vol. 20. pp. 368-380.

Musaka, S. (n.d.), 'Are Expatriate Staff Necessary in International Development NGOs? A Case Study of an International NGO in Uganda'. CVO *International Working Paper* 4.

Mutua, *et al.* (1996), 'The View from The Field: Perspectives from Managers of Microfinance Institutions'. *Journal of International Development.* Vol. 8. No. 2. pp. 179-193.

Najam, A. (1996), 'NGO Accountability: A Conceptual Framework'. *Development Policy Review.* Vol. 14. pp. 339-353.

News from Bangladesh (Dhaka) (1998), 'Madrasahmen Swoop on NGO Activists: B'Baria Clash Hurts 150'. www.Bangla.org/news/amitech. 8 December.

News from Bangladesh (Dhaka) (1997), 'BRAC Female Staff Assaulted, Robbed'. www.Bangla.org/news/amitech. 9 April.

NGO Affairs Bureau (1998), *Flow of Foreign Grant Funds Through NGO Affairs Bureau at a Glance.* Dhaka: NGO Affairs Bureau, PM's Office/GOB.

Nyamugasira, W. (1998), 'NGOs and Advocacy: How Well are The Poor Represented?'. *Development in Practice.* Vol. 8. No. 3. pp. 297-308.

Observer/CAF (1997), *The Usual Causes? How Charities are Shaping up for the Millennium.* London: Observer/CAF.

Odwyer, T. and Woodhouse, T. (1996), 'The Motivations of Irish Third World Development Workers'. *Irish Journal of Psychology.* Vol. 17. No. 1. pp. 23-34.

Olive (Organization Development and Training). (1996), *From Political to Developmental Practice: Facing The Challenges of Fieldworker Development and Training.* Durban: Olive.

Omvedt, G. (1989), 'Class, Caste and Land in India: An Introductory Essay'. In Alavi. H. and Harriss, J. (eds). *South Asia.* London: Macmillan.

O'Neill, M. (1989), *The Third America: The Emergence of The Nonprofit Sector in The United States.* San Francisco: Jossey-Bass Publishers.

Onyx, J. and Maclean, M. (1996), 'Careers in The Third Sector'. *Nonprofit Management and Leadership.* Vol. 6. No. 4. pp. 331-345.

Osborne, S. P. (1996), 'What Kind of Training Does The Voluntary Sector Need?'. In Billis, D. and Harris, M. (eds). *Voluntary Agencies - Challenges of Organisation and Management.* London: Macmillan.

Overseas Development Administative (1995), *Bangladesh-Country Aid Programme 1995.* Dhaka: ODA.

Overseas Development Institute (1995), NGOs and Official Donors. ODI *Briefing Paper.* No. 4.

Palli Karma Sahayak Foundation (PKSF) (1998), *PKSF - A Guideline.* Dhaka: PKSF.

Palmer, P. and Hoe, E. (eds). (1997), *Voluntary Matters Management and Good Practice in The Voluntary Sector*. Norwich: The Directory of Social Change and The Media Trust.

Palmer, T. G. (1997), 'Untitled'. *Civnet*. Vol. 1. No. 2. (from world wide web).

Pastner, C. M. (1982), 'Rethinking the Role of the Women Field Worker in Purdah Societies' *Human Organisation*. Vol. 41. No. 3. pp. 262-264.

Paul, S. (1991), 'Nongovernmental Organisations and the World Bank: An Overview'. In Paul, S. and Israel, A. (eds) *Nongovernmental Organisations and the World Bank - Cooperations for Development*. Washington DC: The World Bank

Pearson, R. (1991), *The Human Resource – Managing People and Work in the 1990s*. London: McGraw-Hill.

People in Aid (2000), 'Agencies Piloting the Code'. www.peopleinaid.org.

People in Aid/Overseas Development Institute (1997), *The People in Aid Code of Best Practice in the Management and Support of Aid Personnel*. London: People in Aid/ODI.

Pratt, B. and Boyden, J. (1985), *The Field Directors' Handbook - An Oxfam Manual for Development Workers*. Oxford: Oxford University Press/Oxfam.

PROSHIKA A Centre for Human Development (1997), *PROSHIKA A Centre for Human Development (Brochure)*. Dhaka: PROSHIKA A Centre for Human Development

Rahman, A. (1999), 'Micro-Credit Initiatives for Equitable and Sustainable Development: Who Pays?'. *World Development*. Vol. 27. No. 1. pp. 67-82.

Rahman, M. M. and Islam, M. N. (1994), 'Role of Service Providers, Programme Managers and Family Planning Workers in the Sterilisation Procedure of Bangladesh'. *Genus*. Jul-Dec. pp. 65-74.

Rangpur Dinajpur Rural Service (RDRS) Bangladesh (1996a), *RDRS Bangladesh 1996*. Dhaka: RDRS Bangladesh.

Rangpur Dinajpur Rural Service (RDRS) Bangladesh (1996b), *Development Programme Policy 1996-2000*. Dhaka: RDRS Bangladesh.

Rao, A. and Kelleher, D. (1995), 'Engendering Organisational Change: The BRAC Case'. *IDS Bulletin*. Vol. 26. No. 3. pp. 69-78.

Rao, A. and Kelleher, D. (1998), 'Gender Lost and Gender Found: BRAC's Gender Quality Action-Learning Programme'. *Development in Practice*. Vol. 8. No. 2. pp. 173-185.

Rashid, M. H. (1994), 'Scenario of Rural Community Resource Centres in Bangladesh'. *Grassroots*. Vol. 4. No. 13-14. pp. 222-226.

Reza, K. (1999), 'Charity Pleasure, Charity Politics'. *Holiday* (Weekend Newspaper). Vol. XXXIV. No. 32. pp.1 & 5.

Riddell, R. and Robinson, M. (1992), 'The Impact of NGO Poverty Alleviation Projects: Results of the Case Study Evaluations'. ODI *Working Paper* No. 68.

Riker, J. V. (1995a), 'From Cooption to Cooperation and Collaboration in Government-NGO Relations: Toward an Enabling Policy Environment for People Centred Development in Asia'. In Heyzer, N. *et al.* (eds). *Government - NGO Relations in Asia - Prospects and Challenges for People-Centred Development*. London: Macmillan.

Riker, J. V. (1995b), 'Contending Perspectives for Interpreting Government-NGO Relations in South and Southeast Asia: Constraints, Challenges and the Search for Common Ground in Rural Development'. In Heyzer, N. *et al.* (eds) *Government-NGO Relations in Asia - Prospects and Challenges for People-Centred Development.* London: Macmillan.

Riker, J. V. (1995c), 'Reflections on Government-NGO Relations in Asia: Prospects and Challenges for People-Centred Development'. In Heyzer, N. *et al.* (eds) *Government-NGO Relations in Asia - Prospects and Challenges for People-Centred Development.* London: Macmillan.

Robinson, M. (1993), 'Governance, Democracy and Conditionality: NGOs and The New Policy Agenda'. In Clayton, A. (ed.). *Governance, Democracy and Conditionality: What Role for NGOs?* Oxford: INTRAC.

Robinson, M. (1997), 'Privatising The Voluntary Sector: NGOs as Public Service Contractors?'. In Edwards, M. and Hulme, D. (eds). *NGOs, States and Donors - Too Close for Comfort?* London: Macmillan.

Rooy, A. V. (1997), 'The Frontiers of Influence: NGO Lobbying at the 1974 World Food Conference: The 1992 Earth Summit and Beyond'. *World Development.* Vol. 25. No. 1. pp. 93-114.

Rostow, W. W. (1960), *The Stages of Economic Growth: A Non-Communist Manifesto.* Cambridge: Cambridge University Press.

Ruebush, T. K. *et al.* (1994), 'Qualities of an Ideal Voluntary Community Malaria Worker: A Comparison of The Opinions of Community Residents and National Malaria Service Staff'. *Social Science Medicine.* Vol. 39. No. 1. pp. 123-131.

Saha, S. (1998), 'Rastrya O Nagarik Samaj' (in Bangla). *Sangbad.* 14 April.

Salamon, L. M. (1992), *America's Nonprofit Sector: A Primer.* New York: The Foundation Centre.

Salamon, L. M. and Anheier, H. K. (1996), *The Emerging Nonprofit Sector - An Overview.* Manchester: Manchester University Press.

Salamon, L. M. and Anheier, H. K. (1997a), 'Introduction: In Search of The Nonprofit Sector'. In Salamon, L. M. and Anheier, H. K. (eds). *Defining The Nonprofit Sector - A Cross-National Analysis.* Manchester: Manchester University Press.

Salamon, L. M. and Anheier, H. K. (1997b), 'Conclusion'. In Salamon, L. M. and Anheier, H. K. (eds). *Defining The Nonprofit Sector - A Cross-National Analysis.* Manchester: Manchester University Press.

Salamon, L. M. and Anheier, H. K. (1997c), 'The Challenges of Definition: Thirteen Realities in Search of a Concept'. In Salamon, L. M. and Anheier, H. K. (eds). *Defining The Nonprofit Sector - A Cross-National Analysis.* Manchester: Manchester University Press.

Salamon, L. M. and Anheier, H. K. (1997d), 'Toward a Common Definition'. In Salamon, L. M. and Anheier, H. K. (eds). *Defining The Nonprofit Sector - A Cross-National Analysis.* Manchester: Manchester University Press.

Salmen, L. F. and Eaves, A. P. (1991), 'Interactions between Nongovernmental Organisations, Governments, and the World Bank: Evidence from Bank Projects'. In Paul, S. and Israel, A. (eds). *Nongovernmental Organisations and the World Bank - Cooperations for Development.* Washington DC: The World Bank.

Save the Children Fund (United Kingdom) (SCF (UK)) Bangladesh (1997), *Bangladesh Country Strategy Paper (Interim).* Dhaka: SCF (UK).

Sayer, A. (1992), *Methods in Social Science: A Realist Approach.* London: Routledge.

Sen, S. (1999), 'Some Aspects of State-NGO Relations in India in The Post-Independence Era'. *Development and Change.* Vol. 30. pp. 327-355.

Senillosa, I. D. (1998), 'A New Age of Social Movements: A Fifth Generation of Non-Governmental Development Organizations in The Making?' *Development in Practice.* Vol. 8. No. 1. pp. 41-53.

Sethi, A. S. and Schuler, R. S. (eds) (1989), *Human Resource Management in the Health Care Sector – A Guide for Administrators and Professionals.* Connecticut: Quorum Books

Schmidt, R. H. and Zeitinger, C. (1996), 'Prospects, Problems and Potential of Credit-Granting NGOs'. *Journal of International Development.* Vol. 8. No. 2. pp. 241-258.

Schuler, R. S. and Huber, V. L. (1993), *Personnel and Human Resource Management.* St. Paul: West Publishing Company.

Schuler, S. R. *et al.* (1997), 'The Influence of Women's Changing Roles and Status in Bangladesh's Fertility Transition: Evidence from a Study of Credit Programmes and Contraceptive Use'. *World Development.* Vol. 25. No. 4. pp. 563-575.

Schuler, S. R. *et al.* (1998), 'Men's Violence Against Women in Rural Bangladesh: Undermined or Exacerbated by Microcredit Programmes?'. *Development in Practice.* Vol. 8. pp. 148-157.

Schuurman, F. J. (1993), 'Modernity, Post-Modernity and the New Social Movements'. In Schuurman, F. J. (ed.). *Beyond the Impasse - New Directions in Development Theory.* London: Zed Books.

Sebahara, P. (1998), 'Reflections on Civil Society'. *The Courier.* No. 170. pp. 95-96.

Seibel, W. (1990), 'Organizational Behaviour and Organizational Function: Toward a Micro-Macro Theory of The Third Sector'. In Anheier, W. and Seibel, H. K (eds). *The Third Sector: Comparative Studies of Nonprofit Organizations.* Berlin: Walter de Gruyter.

Seibel, W. and Anheier, H. K. (1990), 'Sociological and Political Science Approaches to The Third Sector'. In Anheier, W. and Seibel, H. K (eds). *The Third Sector: Comparative Studies of Nonprofit Organizations.* Berlin: Walter de Gruyter.

Siddiqui, K. (1987), *Benefit Packages Received by Functionaries of Government Organisations and NGOs.* Dhaka: ADAB.

Sinha, S. and Matin, I. (1998), 'Informal Credit Transactions of Micro-Credit Borrowers in Rural Bangladesh'. *IDS Bulletin.* Vol. 29. No. 4. pp. 66-80.

Slavin, S. (1988), 'Different Types of Nonprofit Managers'. In O' Neill, M. and Young, D. R. (eds). *Educating Managers of Nonprofit Organizations.* New York: Praeger.

Slim, H. and Thompson, P. (1993), *Listening for a Change - Oral Testimony and Development*. London: Panos.

Smith, B. H. (1993), 'Non-Governmental Organizations in International Development: Trends and Future Research Priorities'. *Voluntas*. Vol. 4. No. 3. pp. 326-344.

Smith, D. H. (1995), 'Some Challenges in Nonprofit and Voluntary Action Research'. *Nonprofit and Voluntary Sector Quarterly*. Vol. 24. No. 2. pp. 99-101.

Streeten, P. (1997), 'Nongovernmental Organizations and Development'. *The Role of NGOs: Charity and Empowerment. The Annals of The American Academy of Political and Social Science*. No. 554. pp. 193-210.

Suzuki, N. (1998), *Inside NGOs - Learning to Manage Conflicts Between Headquarters and Field Offices*. London: IT Publications.

Tendler, J. (1983), 'Ventures in The Informal Sector and How They Worked Out in Brazil'. USAID Evaluation Special Studies No. 12.

Tendler, J. (1989), 'Whatever Happened to Poverty Alleviation?' *World Development*. Vol. 17. No. 7. pp. 1033-1044.

Tetzlaff, R. (1997), 'Democratisation in Africa: Progress and Setbacks on a Difficult Road'. *Development and Cooperation*. No. 4. pp. 8-13.

Townsend, J. G. *et al.* (1995), *Women's Voices from The Rainforest*. London: Routledge.

Tvedt, T. (1998), 'NGOs' Role at 'The End of History': Norwegian Policy and The New Paradigm'. In Hossain, F. and Myllyla, S. (eds) *NGOs Under Challenge: Dynamics and Drawbacks in Development*. Helsinki: Ministry of Foreign Affairs of Finland. pp. 60-83.

Umar, B. (1999), 'Bangladesh: Intellectuals, Culture and the Ruling Class'. *Economic and Political Weekly*. 15 May. pp. 1175-1176.

Unia, P. (1991), 'Social Action Group Strategies in The Indian Sub-Continent'. *Development in Practice*. Vol. 1. No. 2. pp. 84-95.

United National Development Programme (1993), *Human Development Report 1993*. London: OUP.

Uphoff, N. (1993), 'Grassroots Organizations and NGOs in Rural Development: Opportunities with Diminishing States and Expanding Markets'. *World Development*. Vol. 21. No. 4. pp. 607-622.

Uphoff, N. (1995), 'Why NGOs are not the Third Sector: a Sectoral Analysis with Some Thoughts on Accountability, Sustainability and Evaluation'. In Edwards, M. and Hulme, D. (eds) *Non-Governmental Organisations - Performance and Accountability:- Beyond the Magic Bullet*. London: Earthscan Publications.

Van der Heijden, H. (1987), 'The Reconciliation of NGO Autonomy and Operational Effectiveness with Accountability to Donors'. *World Development*. Vol. 15 (supplement). pp. 103-112.

Vivian, J. (1994), 'NGOs and Sustainable Development in Zimbabwe: No Magic Bullets'. *Development and Change*. Vol. 25. pp. 181-209.

Vivian, J. and Maseko, G. (1994), 'NGOs, Participation and Rural Development - Testing The Assumptions with Evidence from Zimbabwe'. UNRISD *Discussion Paper* No. 49. Geneva: UNRISD.

Vladeck, B. C. (1988), 'The Practical Difference in Managing Nonprofits: A Practitioner's Perspective'. In O' Neill, M. and Young, D. R. (eds). *Educating Managers of Nonprofit Organizations.* New York: Praeger.

White, S. (1992), *Arguing with the Crocodile: Gender and Class in Bangladesh.* London: Zed Books.

White, S. (1999), 'NGOs, Civil Society, and the State in Bangladesh: The Politics of Representing The Poor'. *Development and Change.* Vol. 30. pp. 307-326.

Wong *et al.* (1998), 'Experiences and Challenges in Credit and Poverty Alleviation Programmes in Bangladesh: The Case of PROSHIKA'. In Wood, G. and Sharif, I. (eds). *Who Needs Credit? Poverty and Finance in Bangladesh.* London: Zed Books. pp. 145-170.

Wood, G. D. (1994), *Bangladesh - Whose Ideas, Whose Interests?* Dhaka: UPL.

Wood, G. D. (1997), 'States Without Citizens: The Problem of The Franchise State'. In Edwards, M. and Hulme, D. (eds). *NGOs, States and Donors - Too Close for Comfort?* London: Macmillan.

Woolcock, M. (1998), 'Social Theory, Development Policy and Poverty Alleviation: A Historical-Comparative Analysis of Group-Based Banking in Developing Economies'. Ph.D. Dissertation. Department of Sociology, Brown University.

World Bank (1989), *The World Bank Operational Manual.* Washington DC: World Bank.

World Bank (1991a), *How The World Bank Works with Non-Governmental Organisations.* Washington DC: World Bank.

World Bank (1991b), *World Development Report 1991: The Challenge of Development.* Oxford: OUP.

World Bank (1994), *The World Bank and Participation.* Washington DC: World Bank.

World Bank (1995), *Working with NGOs: A Practical Guide to Operational Collaboration Between The World Bank and the NGOs.* Washington DC: World Bank.

World Bank (1996a), *The World Bank's Partnership with Nongovernmental Organizations.* Washington DC: Participation and NGO Group, Poverty and Social Policy Department/The World Bank

World Bank (1996b), *Bangladesh: Labour Market Policies for Higher Employment.* Washington DC: World Bank

World Bank (1996c), *NGOs and The World Bank: Incorporating the FY95 Progress Report on Cooperation between The World Bank and NGOs.* Washington DC: World Bank.

World Bank (1996d), *Bangladesh - Poverty Alleviation Microfinance Project (Staff Appraisal Report).* Washington DC: World Bank.

World Bank (1996e), *Bangladesh - Rural Finance.* Washington DC: World Bank.

World Bank (1996f), *The World Bank Participation Sourcebook.* Washington DC: World Bank.

Zaidi, S. A. (1999), 'NGO Failure and The Need to Bring Back The State'. *Journal of International Development*. Vol. 11. pp. 259-271.

Zaman, H. (1998), 'Can Mis-Targeting be Justified? Insights from BRAC's Microcredit Programme'. *IDS Bulletin*. Vol. 29. No. 4. pp. 59-66.

Zhouri, A. (forthcoming), 'Pathways to The Amazon: British Campaigners in The Brazilian Rainforest'. In Chamberlain, M. and Roberts, K. L. *Environmental Consciousness and Environmental Movements, Memory and Narrative Stories*. Vol. 3. London: Routledge.

Index

Printed and bound by CPI Group (UK) Ltd, Croydon, CR0 4YY

22/10/2024

01777625-0004